The
GOVERNANCE
of KNOWLEDGE

The GOVERNANCE *of* KNOWLEDGE

Nico Stehr, *editor*

Routledge
Taylor & Francis Group
LONDON AND NEW YORK

First published 2004 by Transaction Publishers

Published 2017 by Routledge
2 Park Square, Milton Park, Abingdon, Oxon OX14 4RN
711 Third Avenue, New York, NY 10017, USA

Routledge is an imprint of the Taylor & Francis Group, an informa business

Library of Congress Catalog Number: 2003047312

Library of Congress Cataloging-in-Publication Data

The governance of knowledge / Nico Stehr, editor
    p. cm.
  Includes bibliographical references and index.
  ISBN 0-7658-0172-8 (cloth : alk. paper)
    1. Knowledge. Sociology of. 2. Science—Social aspects. I. Stehr, Nico.

BD175.G65 2003
306.4'2—dc21

                                        2003047312

ISBN-13: 978-1-4128-6409-1 (pbk)
ISBN-13: 978-0-7658-0172-2 (hbk)

Contents

Preface

Anxieties and concerns about the social consequences of new scientific knowledge and novel technologies are not new. Nor are elusive promises of the plain blessings of science for humankind and the mitigation of human suffering that scientific advances entail. But a persuasive case can be made that we have reached a new, modern stage. The first controlled genetic experiment did not occur until 1972. The first human being conceived outside a woman's body was born in 1978. The controversial discussions of recombinant DNA, embryonic stem cells, genetically modified foods, the prospects of lethal and "nonlethal" biotechnology weapons, genetic engineering of the human germline, the neurosciences, the reconstruction of the genome of the ancestor of the human being, neurogenetics, and reproductive cloning exemplify some of the novel issues we are confronting in vigorously contested debates (see also the essay by William Leiss, Chapter 5 in this volume). Concerns about the societal consequences of an unfettered expansion of (natural) scientific knowledge are raised more urgently and are moving to the center of disputes in society and to the top of the political agenda. Governments will have to engage in new political activity and will be held accountable to new standards. Public conflicts, friction, and disputes over the implementation of knowledge, seen by at least some as attacks against science, will no longer mainly take place *a posteriori.*

For it is no longer uncommon that the discussion of the role of new scientific knowledge in modern societies leads to demands that such knowledge and its impact be regulated, managed in some way, or even suppressed. The question has now become pertinent (and these concerns are not merely prompted by the possible consequences of nuclear science and technology anymore), how can we avoid being devoured by a marvelously powerful science? It would be too easy to dismiss calls for intervention as irrational or antimodern responses. The concerned questions being raised and the resistance being mobilized refer back, of course, to the image that we have of ourselves, according to which we are arranging our life—an image that appears to be under threat.

But what knowledge needs to be regulated, and whose responsibility is it? And what knowledge do we require in the context of regulative knowledge politics? Do we need to regulate new scientific knowledge as closely as, or even more strictly than, one regulates, say, traffic? If not, on what grounds? If so, for what reasons, and protecting what interests? If strict restrictions apply, could absolute prohibitions (for example, outlawing human germline gene therapy) be enforced? Is knowledge about to become a private good (again)? What will be the identity of some of the major actors involved, and how will knowledge politics be organized?

We are interested in the bases of the approaches and clashing perspectives in debates about the policing of knowledge, as well as the methods and chances of regulating new knowledge in modern and increasingly fragile societies. Crucial to the possibilities and limits of knowledge politics are dynamic social, cultural, political, and economic settings, often at the macro rather than micro level, such as laboratories, theories, and experiments. The context-specific matrix changes rapidly; for example, as a result of the loss of power of large, modern societal institutions such as the state, the church, corporations, and science. Yet large social institutions continue to present themselves, despite their growing weaknesses, as systems of effective societal control. Robust scenarios for knowledge politics are difficult to generate and are not at all obvious. The burden of this study is nevertheless to show how and why knowledge politics are set to emerge as an arena of significant political activity and choices; and that much can be anticipated. Knowledge politics will have a major impact on what future societies will look like.

Acknowledgments

With a single exception, all the papers included in this anthology were first presented at a conference of the Kulturwissenschaftliche Institut (Center for Advanced Cultural Studies) in Essen, Germany, 5–7 September 2001. I am very grateful to the Institut, its director, Professor Jörn Rüsen, for inviting me to become a fellow of the Kulturwissenschaftliche Institut, and for supporting the international conference on "The Governance of Knowledge." I am indebted to Professor Michael Polanyi of the University of Toronto who opened the conference in Essen with a stimulating discussion of the "Governance of Science." My sincere thanks also go to Dr. Norbert Jegelka, the executive director of the Institut, his great enthusiasm and usual facility to smooth administrative hurdles. We all benefited that Doris Alemara of the Institut made sure that the conference was well organized in all its many details and that the participants felt at home in the Kulturwissenschaft-liche Institut. Finally, I have to thank Robin Taylor who took it upon herself to copyedit the manuscript with her familiar efficiency and care.

Nico Stehr

Introduction

A World Made of Knowledge

Nico Stehr

Will what can be shown always be done?[1] This anthology is an initial response to this question, and it can only be an interim report. Using a variety of perspectives and approaches, we plan to discuss what we understand may well become one of the most significant and contentious issues for intellectual, legal, public, scientific, and political discourse and action during the century that has just begun:[2] the growing pressure to regulate or police *novel* knowledge—in other words, the emergence of a new field of political activity, namely knowledge politics and policies.[3] What is at stake is more than merely the vague feeling that a slowdown or a consolidation in the fabrication of the volume of new knowledge is in order.

Knowledge Politics

Knowledge politics, or governance of knowledge, as a new field of political activity and policy is about attempts to deliberately channel the social role of knowledge, to generate rules and enforce sanctions pertaining to relevant actors and organizations, to affix certain attributes (such as property restrictions) to knowledge and—likely the most controversial strategy—to restrict the application of new knowledge and technical artifacts. For the main part the governance of knowledge will consist of political, economic, and legal means and efforts located for the most part *outside* the immediate boundaries of the scientific community. Scientists will of course play a central role in any such efforts. As a matter of fact, Gemot Böhme, in his essay on knowledge policy (Chapter 1 in this volume), argues that the task of knowledge politics ought to be centered in science itself and therefore become a core issue for the scientific community.[4]

The essence of knowledge politics consists of strategic efforts to move new scientific and technical knowledge, and thereby the future, into the center of the cultural, economic, and political matrix of society.[5] Despite what the term may appear to suggest on the surface, knowledge politics are not inherently prohibitive. The regulation of knowledge—in the general sense of attempts to control it, and not merely through statutory enactments and administrative regulations and decisions—also extends to efforts designed to enhance and enlarge the options and opportunities for the use of new knowledge in society. Many segments of civil society want governments to enact measures aimed at satisfying their ideas or demands for the future use of knowledge, be it in the field of health care, education, the environment, or social policies in varied fields within and across nations. As a matter of fact, as long as at least some societies continue to place a strong emphasis on the virtue of individual initiative and decisions based on self-interest, modern social institutions, if only passively, encourage the unencumbered use of novel scientific knowledge and technical devices (Green 1976: 171).[6]

However, in this anthology we will focus on the control of knowledge (which suggests that knowledge generated by molecular genetics, for example, may involve "adverse" individual or collective consequences and that a "policing" of knowledge may be justified on these grounds), rather than the extension of ways of deploying new knowledge more extensively in modern society. Although one should not underestimate the persistence of governance regimes once institutionalized, an analysis of the governance of knowledge in modern society has to be cognizant of the general practical incompleteness, fragility, obsolescence, and often failure of projects aimed at governance in modern societies, the variety of institutions likely involved (see also the essay by Werner Rammert Chapter 4 in this volume)[7] as well as, more narrowly, the possibility that rapid deregulation follows on the heels of regulatory efforts (as was the case in the field of genetic engineering in the early 1980s). The tempo with which knowledge evolves may be a further reason why specific attempts to regulate knowledge may become obsolete. Regulative politics is surpassed by the dynamics of knowledge.[8]

The Emergence of Knowledge Politics

Why are knowledge politics emerging? Why the growing efforts to exert power over knowledge? Why are we, perhaps in growing numbers, not prepared to simply accept the apparently "natural" progression, to take for granted the relentless, exponential development of scientific knowledge, as well as technical artifacts and their application, as a key to unlocking the mysteries of the world, as a release from pain and freedom from suffering, as the basis for a better and just society, as a means for greater prosperity; or to believe that more knowledge simply represents the master key to an emancipation from all kinds of troubling ills and harsh constraints? The straightforward, or at least traditional, assumption that specialized knowledge ought to command respect in general, and that any increase in knowledge automatically brings with it an increase in benefits to humankind in particular, is becoming porous and vulnerable.[9] The idea that the uselessness of science is a virtue and that the uses that humans "have drawn from science have contributed to their misery" (Chargaff 1975: 21) is but a marginal voice, rarely heard even today. The optimistic faith, nurtured in the West immediately after WWII and then in a period of unprecedented economic growth in the 1950s and early 1960s, that a constant expansion of "knowledge" might even prompt a displacement of politics and ideology (Bell 1960, 1973; Brooks 1965; Lane 1966) has been thoroughly demystified (King and Melanson 1972).

If one no longer regards the fabrication and use of additional scientific knowledge as a humanitarian project, "as an unquestioned ultimate good, one is willing to consider its disciplined direction" (Sinsheimer 1978: 23). The fear that we know too much and that we are about to assume the role of God (or that we are about to engage in a "self-transformation of the species"; Habermas 2001: 42) increasingly replaces the concern that we do not know enough and that we are to a large degree poorly informed. Apprehension and alarm replace the rhetoric of hope that until recently dominated societal discourse about new developments in science and technology in modern societies (compare Mulkay 1993: 735–9, as well as Kevin Jones' contribution about the recent BSE crisis in England. Chapter 8 in this volume).

The concern that we know too much is no longer—as it was in the 1970s, for example—that we are amassing a large store of trivial and practically irrelevant knowledge at a high price with no promise of utility or gain (Lübbe 1977: 4). It has been replaced by concerns about the accumulation of novel knowledge that appears to have questionable social consequences. In that sense, current concerns directed toward science represent a return to vigorous social conflicts science has experienced in the past. But in contrast to past disputes, when discussions of the societal consequences of science were driven by complaints about its deficient social and economic utility in tackling major social problems of the day (e.g., war, crime, unemployment, poor educational standards, family breakdown), today concern is increasingly directed toward a surplus of effects—especially with respect to traditional world views, the established life-worlds, and limits to what may be manipulated in nature and society.

More specifically, the economic, political, legal, and societal importance of knowledge policy in this century escalates because of the following.

1. We are faced with *new forms of knowledge* (or new types of capacities to act)[10] that both set off alarms and encourage broad sets of promises. The route from basic research to applied research to commercial application is in some fields of science, such as molecular biology, a particularly short and direct one.[11] The difference between basic and applied research diminishes. The identification of a gene constitutes the test for a particular gene. The transformation of knowledge is co-determined by an increasing specialization in science and a massive infusion of private and public funds in support of particular research fields. The limits of what is feasible in practice are decisively displaced. Knowledge itself is transformed. Knowledge now emerging is more powerful in shifting or destroying the boundaries of the possible; for example, the techniques of genetic engineering make DNA subject to direct human access and control.[12] The bio-utopian future, as it is promised, amounts to a control of the biological destiny of mankind. With the emancipation from "their genetic straightjackets and the constraints of Nature, people will find a new land of freedom—a freedom to overcome disease and hunger, to have an improved standard of health and, of course, to live longer" (Hindmarsh, Lawrence, and Norton 1998: 5).

2. As the result of the *speed* with which the *volume* of the available stock of knowledge grows, the opportunities of *contacts* with knowledge, including its adverse impacts, multiply.[13] As a matter of fact, the very definition of and concern with what constitutes the speed of cultural evolution invariably refers to the presence and accelerated appearance of incremental knowledge.[14]

3. The rapid increase of knowledge multiplies, by definition, our *capacities to act*; for knowledge represents capacities of action or models for reality that considerably enlarge our options for changing social realities. The economic, political, and social centrality of knowledge grows. For the political system, constantly on the lookout for political topics, new scientific knowledge constitutes problems that have to be dealt with politically. The enlarged capacities to act bring about an increase in concerns about the various effects of, as well as access to, control, distribution, benefits, and costs of new knowledge (see Horowitz 1985). Nothing appears to be impossible

anymore. The prevailing public sentiments about the benefits and costs associated with our enlarged ability to alter the environment and society are changing.

4. Although every past technical invention and scientific breakthrough has produced *responses* of exhilaration, as well as deep concern and dire predictions about its social or psychological impact, there is a tendency to shift when it comes to the assessment of the role of science and technology in society, from a willingness to engage in *a posteriori* cleanup towards efforts to reduce or even prevent harmful effects. But questions such as "How can we possibly use it?" invariably compete, at least in the field of medical inventions, with questions such as "How could we not?" The rapidity with which incremental knowledge is produced has not only increased the awareness that knowledge becomes the motor of social change, but has also heightened the sense of alarm, risk, and *uncertainty* that are seen to be associated with the transformative capacity of knowledge. Similarly, even an element of outright hostility may be one of the visible responses from the general public—for example, with respect to the heightened tempo of scientific and technological "progress" and its anticipated effects. During periods of accelerated social change, demands for planning, regulating, and policing the forces of change always grow as well. Rapid social change generated by knowledge is no exception.

5. Efforts to regulate and control incremental knowledge cannot be uncoupled from *time* and *place.* As a matter of fact, the importance of the context and the boundaries of the context within which efforts may be launched to control knowledge immediately point to one of the dilemmas knowledge policy inevitably faces, even in a world that is supposedly shrinking as the result of the forces of globalization: namely the limits of control, legitimacy, and authority to police knowledge across contingent boundaries and borders.[15]

6. Finally, the extensive industrial, military, and medical application of science in the last century accounts for the kind of esteem it widely enjoys among the citizens and institutions in modern society. Scientism, or technocratic perspectives that deny the possibility of alternative decisions about the use of science and technology, take the repute of science to the extreme. The "authority of science" is virtually impossible to doubt. However, the taken-for-granted social standing science and scientific knowledge enjoyed in modern societies—for example, as a rather neutral, value-free, instrumental agency—has never been without its detractors. Scientism is under increasing scrutiny. Widespread scrutiny of science and technology dating at least from the 1960s, and concerns voiced then about the environmental impact of science-based products, allows for the possibility of proposing and justifying limits to the deployment of novel scientific knowledge in society.

We will try to offer some realistic answers to these and associated questions about the regulation of knowledge in modern societies, such as the possibilities, foundations, prospects, and effectiveness, but also the limits of (modern) knowledge politics in an increasingly globalized world. The answers found in this volume use both case studies (e.g., the linkages and embeddedness of knowledge production and politics in the case of deploying race as a category as examined by Troy Duster in Chapter 7; Javier Lezaun's discussion of the controversial debates surrounding genetically modified foods in Chapter 9)[16] or different theoretical approaches (such as Wolfgang van den Daele's observations about the role of traditional knowledge in modern societies; Chapter 2).

Moreover, one can ask, would knowledge permitting an extension of the average human life expectancy not be applied almost instantly after it had been discovered as a capacity for action? Once medical intervention is possible prior to the onset of a disorder, why wait until someone falls ill? But should we not fear, on the other hand, a much improved predictability of individual life expectancy or therapy preceding an illness? Might such predictability of the life span of the individual not eliminate much of the spontaneity of action or lead to horrible mistakes? Do we want to live in a world in which control of all conquerable genetic defects is possible? In what ways will the state or other corporate actors intervene between prospective parents and their ability to decide about the genetic makeup of their children? Should the prerogative of individual autonomy prevail in these cases, or should collective prerogatives govern decision-making about how to approach the potential use of new knowledge? The legend of Faust resonates precisely with the ambivalent difference that accompanies the fascination of a persistent quest by science for missing answers and the fear about what such a pursuit of the unknown may engender.

All of these issues become even more interesting and perplexing in light of the observation that we are in fact living in an age of deregulation, or that those who advocate the withdrawal of the state by pushing a neoliberal policy agenda have won the day. At least within the developed world, there appears to be no exception to the strong support for neoliberal policies that promote deregulation efforts, be it by freeing labor markets, by lowering taxes, or by withdrawing from strong welfare-state policies (Cerny 1991).

The politics of regulating new knowledge and novel technical devices is bound to upset the established line of political conflicts, and in many instances may well create "strange political bedfellows" in the form of novel and quickly changing political coalitions. Central emotionally and politically charged debates in modern society about the authority of science, medicine, or experts, but also about politics and the control of the body, the desirable relations between nature and society, the meaning of technology and human agency; the linkages between ethics and knowledge, will not only be symbolically recast and heavily strained, but also re-invented.[17]

Science and Knowledge Policies

It is necessary from the outset to refer to a dual difference that ought to be sustained in a discussion of the governance of knowledge, as far as is possible, though this may be difficult in some instances: namely, the difference between science and knowledge policy. The focus in this anthology is on knowledge, not on science policies. Thus, the contested issue of human embryonic cell research, much in the public eye in the last few months, is of course a question that goes to the heart of science policy.[18] Science policy as conducted by governments, firms, and foundations refers directly to the constitution of scientific knowledge, the individuals who produce such knowledge, the social context within which knowledge is fabricated, incentives such as tax policies, tariffs, subsidized R&D, and the alleged benefits of science for society; which legitimate various efforts to "manage" science. Science and technology policies are drawn up and implemented using a variety of instruments, some of which I have just listed, in social systems outside the scientific community. The goal, of course, is to gain leverage on the fabrication of scientific knowledge and the development of technologies.

In contrast, the capacities to act on the knowledge generated by research on (embryonic) stem cells extracted from human embryos (which have the potential to grow into any cell or tissue in the human body and therefore might be instrumental, as its proponents suggest, in curing degenerative diseases such as Parkinson's, Alzheimer's, heart disease, kidney failure and diabetes) pertains to the emerging field of knowledge politics. Many diseases, as the U.S. National Bioethics Advisory Commission (1999: 20) emphasizes, such as Parkinson's disease and juvenile-onset diabetes mellitus, are triggered as a result of the death or dysfunction of just one or a few cell types. A substitution for dysfunctional cells could offer effective treatment and even cures for such illnesses. The promise that research efforts using embryonic stem cells are likely to yield considerable therapeutic benefits indicates already that a liberal mix of science and knowledge policy assertions occurs in such disputes.[19] One set of the assertions under dispute refers to science policy matters, such as resource allocation, while others already anticipate new knowledge claims and instrumental abilities that become central to discourse in knowledge policy discussions. A bill passed by the U.S. House of Representatives in late July 2001 that prohibits human cloning ("reproductive cloning") and the cloning of embryonic stem cells ("therapeutic cloning")[20] directly affects research in this field.[21] None the less, it is entirely possible that knowledge policies become indistinguishable from research policies as intentions and agendas of the former extend into and intervene directly or indirectly in the production of knowledge in the scientific community. A less tangled example of knowledge, in contrast to science, policy would be the curtailment in 1975 by the government of a Harvard-based genetic screening program for XYY chromosome patterns. The genetic work, using known techniques, was controversial because it pursued the idea that there was a significant correlation between deviant behavior and the presence of the XYY chromosome. Pressure from the Children's Defense Fund and similar groups brought about a ban enacted by the Reagan administration.

In addition, the agenda building and the nature of specific science and knowledge policies likely depend on common sociopolitical convictions or trends, such as the resolve either to pursue strongly interventionist science and knowledge policies or to support and strengthen the autonomy of the scientific community and market forces. The discursive, heuristic distinction between science and knowledge policy does not imply that new knowledge is, in the context of societal change, the outcome of an exogenous process. Novel knowledge results from endogenous social processes. It does not fall from the sky or appear by accident on the scene.

The Regulation of Embedded and New Forms of Knowledge

A second immediately relevant distinction helpful in delineating the issues at hand refers to the regulation of existing forms of knowledge and the regulation of additional knowledge that still has to be realized. Culture, in a most general sense, constitutes control. Culture dictates and regulates. And it has done so from the beginning. What is new is the tempo with which new knowledge is generated—additional knowledge that needs to be assessed and controlled in some fashion, at least in the eyes of a growing segment of the public in modern societies, and that needs to be brought to the market quickly, at least in the view of states, corporations, governments, and individuals anticipating certain benefits from the application of the new knowledge.

I refer to but a few relevant announcements among the growing number of recent news items that can be deployed to illustrate the issue of knowledge policy: In September of 2001, the acting head of the ethics committee of the American Society for Reproductive Medicine announced that it is sometimes acceptable for couples to choose the sex of their children by selecting cither female or male embryos, discarding the rest. One set of fertility clinics was quick to respond. The chairman of the board of the Center for Human Reproduction indicated that they would offer the procedure immediately.[22] The U.S. Food and Drug Administration indicated in early June 2001 that it does not want meat or milk from cloned livestock sold to consumers. First it has to be established that the food is "safe" and that the technology does not "harm" the environment or the animals. A study carried out by the National Academy of Sciences is supposed to provide answers to these questions (*International Herald Tribune.* "Cloned Livestock Subject of Review," 6 June 2001: 3). Also in June, the German ministry of consumer affairs and agriculture announced that a particular brand of genetically engineered com seed (Artius) would not be permitted to be used for commercial purposes on German farms. The seed in question is resistant against a particular herbicide. If that herbicide is used, it only destroys other plants and leaves the corn unaffected. Further studies by a state research institute are required. The genetically engineered brand of com seed would have been the first such seed that would have been permitted to be sold and freely used (*Frankfurter Allgemeine Zeitung,* "Zulassung von Genmais vorerst gestoppt," 6 June 2001: 4).[23]

An unambiguous observation about the societal regulation of power (in the Weberian sense of the term) by John Kenneth Galbraith offers the following proposition: "The precision and effectiveness of the regulation of the use of condign power are, perhaps, the clearest index of the level of civilization in a community, and they are extensively so regarded in practice" (1983: 83). If this is the case and, as I will assert, among the growing sources of power in modern society is new or additional knowledge, then by analogy the regulation of the use of such knowledge becomes an indicator of the civility of social relations in modern society. Knowledge politics will be a strongly contested form of regulative politics. But that there will be knowledge politics is a certainty. We should not have any excessive hopes, however, that our ability to anticipate (in any robust sense) the social impact of the use of novel capacities to act (knowledge) will be very impressive. Similarly, knowledge politics will be enacted even though the ability to forecast the consequences of intervention in systems *other* than the political system is likely quite limited (Luhmann [1991] 1993: 155).

Thus, it is important to reflect more directly and intensively about the kind of knowledge we need as well as about the use we make of the knowledge we have. As I have emphasized, we will concentrate in this anthology on the importance of the use of incremental, new knowledge in society. Such a focus extends, of course, to what is now sometimes called "biopolitics," in as much as the term suggests that it is the function of the political system to regulate nothing less than the way we deal with the nature of life, especially in response to new capacities that would affect the ways this system could act on our life and that of others.

The emergence of knowledge politics occurs with some delay in response to the exceptional growth and speed with which knowledge and technical capacities are added in modern societies. Appropriating Adolph Lowe's astute insights (1971: 563), it is a change from social realities in which "things" simply "happened" (at least from the point of view of most people) to a social world in which more and more things are "made" to happen. Advanced society may be described as a knowledge society because of the penetration of all its spheres by scientific and technical knowledge. In knowledge societies, the individual's capability of what she/he can do and be is considerably enhanced. The societal changes I have in mind can also be described in the following way: in the case of large and influential social institutions, but also in the case of individuals and small social groups, the weight in the relation of autonomy and conditionality is shifting. The sum total of conditionality and autonomy is not constant. Both autonomy and conditionality of social action are capable of growing. They may also decline. In knowledge societies, the degree of apprehended autonomy of individuals and small social groups increases, while the extent of conditionality shrinks. In the case of large collectivities such as the state, large corporations, science, the church, etc., the extent to which their conduct is conditioned may decline as well, but their autonomy or ability to impose their will does not increase in proportion. While the limits of what can be done are re-written, the responsibility for the changes that are underway must be shared by a larger segments of society.[24]

The boundaries of what at one time appeared to be solidly beyond the ability of all of us to change, to alter or manage are rapidly moved and penetrated. This applies, for example, to the possibility that we may come to review the validity of Lamarckian ideas that deliberately induced genetic transformations in one individual may in fact be passed to one's offspring in the not too distant future. The result, of course, is that new knowledge and new technical abilities as capacities to act are also perceived as a peril posed to every woman, man, and child, as a threat and a burden not merely to privacy, the status quo, the understanding of what life is, and the course of life, but also as a danger to the very nature of creation. For as the biologist Robert Sinsheimer put it shortly after the discovery of the possibility of genetic engineering by recombinant DNA techniques:

> With the advent of synthetic biology we leave the security of that web of natural evolution that, blindly and strangely, bore us and all of our fellow creatures. With each step we will be increasingly on our own. The invention and introduction of new self-reproducing, living forms may well be irreversible. How do we prevent grievous missteps, inherently unretractable? Can we in truth foresee the consequences, near- and long-term, of our interventions? By our wits mankind has become the master of the extant living world. Will short-sighted ingenuity now- spawn new competitors to bedevil us? (Sinsheimer 1976: 599)

The growth of knowledge and technical capacities is not merely prompted by sheer curiosity interested in penetrating the secrets of nature and society, but also driven by economic and military interests. In deploying novel knowledge and technical artifacts for economic growth and military purposes, the social costs and environmental burdens produced are treated as exogenous and ex-post-developments. As the term "exogenous costs" signals, perceived burdens and costs are mitigated as far as possible only after the realization of new knowledge. A growing gap between perceived benefits and burdens will, of course, enhance calls for the proactive regulation of new knowledge and technical capacities. Vanderburg, for example, refers to the existence of a "labyrinth of technology" in modern societies (2000: xi); that is, the extent to which these civilizations are trapped within the dilemma of first creating burdens of various kinds as the result of making use of science and technology and then mitigating these costs. The labyrinth of technology calls for the "creation of an approach for the engineering, management and regulation of modern technology that proactively prevents social and environmental burdens."

Prospects

The massive difference in and additions to human capacities to act within just a century may well be represented by two "bookmarks": In 1945 humans had produced the capacity to destroy life on earth on a grand scale, while by 2045 it might be possible to create life on a grand as well as a minute scale (see Baldi 2001: 163). Thus, it seems, the speed with which new capacities to act are generated forces us to alter our conceptions of who we are and, even more consequentially, may in fact change who we are. The promises and anxieties raised by these prospects are the motor of knowledge politics in modern societies. The boundaries of what was once clearly beyond the control of all of us are rapidly shifting.

The political landscape is changing as a result of new scientific discoveries and new technological innovations. The kind of regulative knowledge politics now in demand is new. Present mechanisms and institutions are unprepared to cope. But governments will be forced to face up to new problems and novel standards: They will have to develop new rules and they will be judged as to whether they are successful in meeting new goals. The nation-state will continue to be of consequence, but less so as an autonomous corporate actor that shapes knowledge politics. The knowledge politics of the nation-state will frequently have to or (in the case of deliberate policy transfers) desire to enact policies of wider global institutions, international treaties, and social movements. However, and this can also already be detected, the tempo with which solutions to the new problems are found will be far outdistanced by the accumulation of new political challenges.

With the advent and the ascent of knowledge politics on the political ladder of importance, one can expect that the general lack of attention the scientific community has paid to what is done with their discoveries in society will also change. The scientific detachment and autonomy that began as a useful barrier against threats to the unencumbered, single-minded pursuit of knowledge will increasingly be seen as an isolating boundary and challenged by the consequences of knowledge politics for scientific work. The atomic bomb, of course, shattered the isolation of science first. But the new capacities of knowledge, and their apprehended impact on individual and society and enlargement of what is possible, will be an equally powerful force that should transform the relations between science and society and the social engagement of scientists. Current public debates already demonstrate that in increasing numbers, scientists are leaving their laboratories and studies in order to take part in political debates about the future of science and the social consequences of scientific developments.

As I have emphasized, efforts to regulate knowledge are not new, but a strong case can be made that we are about to reach a new stage in the next few decades in efforts by various corporate actors to police new knowledge. Past legal and regulatory practices will probably prove to be of limited value as a guide and precedent for future practices. The disputes, debates, and dilemmas over what discourse (e.g., political, normative, military, or economic considerations) should be decisive in decisions that draft and enact knowledge policies are bound to escalate.

Notes

1. In a report in *Time* (21 July 2002) about ongoing efforts of the Pentagon to develop future "nonlethal" weapons such as "Webs and nets," "maloderants," and "drugs and bugs" (such as "bioengineered bacteria that are capable of eating asphalt, fuel and armor body, or faster acting, weaponized forms of antidepressants, opiates and so-called 'club drugs' that could be administered to unruly crowds") the authors conclude with the admonition "maybe some science fiction should remain fictional."

2. Reference should be made here to Marlin Schulte's discussion of the use of scientific knowledge in the legal decisions (Chapter 10 in this volume) or to Lawrence Lessig's examination of the "innovation commons" (Chapter 11).

3. A report issued by the Rand Corporation (Fukuyama and Wagner 2000: 1; see also Fukuyama 2002) anticipates in an analogous sense that in the early part "of the twenty-first century, the technologies emerging from the information and biotechnology revolutions will present unprecedented governance challenges to national and international political systems." The report deals the governance of both research and knowledge policies. Harriet Zuckerman (1986: 342) has also written about the need to critically examine the use that is made of the knowledge we have.

4. In Chapter 12, Stephen Turner examines the role of exports, politicians, and the state in the governance of knowledge in comparative perspective and poses the puzzling question, "if those who are empowered to make decisions about science cannot understand the science they are making decisions about, or perhaps even meaningfully communicate about it, is the intelligent exercise of popular control over science possible?" (Chapter 11).

5. A discussion of the economics of biotechnology that can no longer be uncoupled from ethical, environmental, legal, and political considerations may be found in Gaisford et al. (2001).

6. Taking the opposite position, Niklas Luhmann ([1991] 1993: 173) observes, perhaps paradoxically reflecting an "old European" political perspective

and experience he always was at pains in opposing, that the pressure to act constitutes the political system: "Politics presents itself as a system of societal control. This alone may dispose it to action rather than inaction. We seldom find mere inaction entered on the credit side of governmental balance sheets."

7. Compare Steve Fuller's contribution (Chapter 3) and his discussion of the role of modern universities in knowledge politics.

8. A case in point could be the discovery of a highly versatile group of adult stem cells isolated from the bone marrow that may make much of the discussion of the use of techniques that rely on embryonic stem cells obsolete (see "Scientists herald a versatile adult cell," *New York Times*, 25 January 2002).

9. Now that the idea that progress in science invariably must equal progress in human affairs is no longer widely taken for granted, it is worth briefly mentioning that this equation that has fallen into disrepute did not find strong support until the age of Francis Bacon, who did much to expedite its acceptance and promotion. Bacon was quite cognizant that the identification of advances in science and society could not be taken for granted but had to be secured first, especially among the ruling classes. In much of antiquity the idea of progress was completely absent from public discourse and in the Middle Ages, human progress was not expected to arrive with or derive from advances in the secular sciences (cf. Böhme 1992).

10. A much more extensive examination of the notion of knowledge as a capacity to act may be found in Stehr (2001).

11. The field of molecular biology, the parent discipline of biotechnology, was until 1972 (when the possibility of genetic engineering by recombinant DNA techniques was discovered) almost without exception an academic field of research with few, if any, practical applications (see Wright 1986).

12. The reference to "techniques of genetic engineering" raises the question whether the differences in the contents of science and technology, otherwise strictly defended (e.g., Pavitt 1987), are in fact vanishing and whether they are still relevant for an examination of knowledge politics. At least in some fields, such as molecular biology, biochemistry and solid state physics, as some would observe (e.g., Dasgupta and David 1986), the boundaries between science and technology are increasingly blurred. Pavitt (1987: 187) acknowledges that in the case of biotechnology patents and biomedical research, science and technology are very close. In other fields, technology largely builds on technology. In general, we are facing what might be called a moving target, and attempts to fix the boundaries between science and technology are bound to be surpassed by practical developments that do not allow themselves to be constrained by terminological debates.

13. The case studies found in this volume by Kevin Jones on the BSE crisis (Chapter 8) and Javier Lezaun on CJM foods (Chapter 9) exemplify such novel development in the relation between science and society.

14. The interim nature of the answers found in this anthology on the emerging issue of the governance of knowledge is at least partly an outcome of the speedup and the unprecedented magnitude with which new technical artifacts are developed and moved to market and new forms of knowledge are fabricated. As a consequence, many of the discussions of and disputes over the social impact of novel knowledge claims can now be found in the media,

especially in daily or weekly publications that are at least able to respond in a more timely fashion to rapid developments. For this reason, we will make more frequent reference than is customary in studies of this kind to news items, comments and contributions found in the "fast" press, rather than in conventional academic journals or books. One also has to be cognizant of the possibility that there may be a deep rift between the knowledge-based future scenarios discussed in the fast media and developments that can in fact be realized in practice.

15. I do not see the emergence of modern knowledge politics as evidence for a backlash against the dominance of "rationalist" thinking and discourse in science as well as mundane reasoning, and therefore as a resurgence of traditional thought, such as "magic, ritual, and superstition," in modern institutions.

16. More specifically, in his paper, Troy Duster is trying to make the case "how and why a 'purging science of race'—where race and ethnic classifications are embedded in the routine collection and analysis of data (from oncology to epidemiology, from hematology to social anthropology, from genetics to sociology)— is neither practicable, possible, nor even desirable. Rather, our task should be to recognize, engage, and clarify the complexity of the inter-action between *any* taxonomies of race and biological, neurophysiological, social, and health outcomes."

17. These brief considerations already point to various limits and limitations of the governance of knowledge. Werner Rammert (Chapter 4) suggests that the possible constraints on knowledge politics "are rooted mainly in two paradoxical processes. First, the heterogeneity of the enrolled actors—scientists and managers, politicians and administrators, venture capitalists and ecological activists—and the diversity of their perspectives cause problems of a successful concertation (Rammert 2000) that does not level out the creative differences between disciplines or institutional rationality standards, and that does not destroy the complementary competences of functionally specialized actors. Secondly, the specificity of knowledge to be an intangible asset and an incompletely explicable set of competences sets limits to the complete control and commercialization of knowledge."

18. Since 1995, the U.S. Congress has every year attached a ban to its appro-priations legislation disallowing support of stem cell research with public monies. Research funds raised from private sources may be used for stem cell research in the United States. As a result, the discoverer of the method to isolate embryonic stem cells (from frozen embryos that were no longer wanted by couples who created them to have children), James A. Thompson of the (Slate) University of Wisconsin, carries out his research on stem cells in a separate laboratory supported by private research funds (see "Scientist's stem cell work creates uproar," *New York Times*, 10 July 2001). The type of political considerations (see Ralph Brave, "Governing the genome," *The Nation*, 10 December 2001) that play a role in decisions for or against public support for research on embryonic stem cells can be gleaned from a report m the *New York Times* ("Bush aides seek compromise on embryonic cell research," 4 July 2001). The decision for or against stem cell research with the aid of federal funds will also be carefully judged—by presidential advisors and observers—against the background of the American public's declining

support of President George W. Bush during his first months in office in the summer of 2001, as reflected in surveys, mainly because the President was seen as too conservative. The presidential advisors have argued against making research funds available. But due also to the low approval ratings of the president prior to 11 September 2001, the political decision has been postponed and newly debated. Whatever the decision will be, some parts of the American voting public will be disappointed.

19. A report issued by the National Institutes of Health (NIH) in June 2001 praises stem cell research and promises a "dazzling array" of treatments of various diseases that at present defy therapy (see "U.S. study hails stem cells' promise," *New York Times*, 27 June 2001). The report not only praises the almost limitless benefits of stem cell research but advocates its more or less unrestricted practice supported by federal funds without analyzing ethical, legal, or social issues (see also NHL "Institutes and Centers answers to the question. 'What would you hope to achieve from human pluripotent stem cell research,'" 26 April 2000, www.nih.gov/ncws/stemcell/achieve.htm).

20. Lanza et al. (2000: 3175), who attempt to make an ethical case for therapeutic cloning, suggest that the term therapeutic cloning, although widely used, is misleading because it "brings to mind images of the replication of a single genome for reproductive purposes. In therapeutic cloning, however, no such replication is involved." The description Lanza and his colleagues advance refers to therapeutic cloning as a new biomedical technology that "involves the transfer of the nucleus from one of the patient's cells into an enucleated donor oocyte for the purpose of making medically useful and immunologically compatible cells and tissues." Both human reproductive cloning (duplicating an entire human organism) and therapeutic cloning currently begin by creating a human embryo.

21. Therapeutic cloning (or cell replacement by means of nuclear transfer) implies that embryos are created in order to experiment with them or use them to treat disease. In Great Britain, therapeutic cloning is legal (but not in the United States, see "House backs ban on human cloning for any objective," *New York Times*, 1 August 2001). President Bush has urged the U.S. Senate to follow the lead of the House of Representatives and ban cloning. The restrictions on "cloning research" prompt some scientists to warn that *limiting science* is "cutting off hope," namely the hope of patients to obtain relief from advances in research in this held. The controversial issue of cloning human embryos has been reignited as the result of the announcement on 25 November 2001 by scientists of Advanced Cell Technology (ACT), a privately held company in Worcester, Massachusetts, that they had used cloning techniques to make several early human embryos, each from a cell taken from an adult (see "Mass, firm's disclosure renews cloning debate," *Washington Post*, 27 November 2001). The company made it clear that the cloned embryos were solely produced for research purposes and not to produce adults. Within 14 days, the embryos, activated oocytes, would be destroyed in the process. During a 4 December 2001 U.S. Senate hearing, the CEO and the chief ethicist of ACT (a professor of religion at Dartmouth) told a subcommittee that life does not begin until two weeks after conception or, in the case of the cloning process, the timing of the nuclear transfer. Why did the representatives of ACT chose the two-week

period? At that juncture, known as the gastrulation or the formation of the primitive streak, the embryo "clarifies" whether it will become a single or two persons. Gastrulation takes place about two weeks after conception. At that point, the embryonic mass organizes or aligns itself into layers. The layers form the first outline of an organism or, in the case of twins, two organisms. The CEO of ACT, Michael West, explained: "If the proposition was that we would clone a developing human being, I would [agree] with Sen. Brownback we shouldn't cross that line. We have a line here. It's the primitive streak . . . Primitive streak. I think, is an effective line to draw and say that is the beginning of a human being, and prior to primitive streak we should use some other terminology . . . because this is not an individualized human being" (as cited in William Saletan, "Everyone's a twinner," *Slate*, 27 December 2001; see also Lanza et al. 2000). In his article, Saletan points to the logical flaws in the position of the proponents of therapeutic cloning and the specific cut-off point they have identified that accounts for the beginning of human life. The flaw involves the accomplished cloning itself, for it represents evidence that the possibility of more than a single person emerging from a particular embryo is not finally determined at the two-week stage of the growth of the embryo. The (later) clone would be another person. The debate and the public relations struggle over cloning are by no means closed. It will become one of the central issues for knowledge politics. The dispute is bound to intensify in the future (see "Dispute over cloning experiments intensifies," *New York Times*, 6 March 2002); but *The Economist* already despairingly comments, "many ethicists . . . are paid consultants to the growing biomedical industry, and not a few scientists in this area are religious believers with a doctrinal agenda" ("Should we lock the door on cell science?" 14 March 2002).

22. The ruling refers to selection procedures using preimplantation genetic diagnostics, and not other techniques (compare "Fertility ethics authority approves sex selection." *New York Times*, 29 September 2001). The September 2001 decision of the ethics committee of the American Society for Reproductive Medicine was revised six months later. The Society now, in a more restrictive opinion, suggests that couples should be discouraged from creating embryos using preimplantation genetic diagnostics, selecting some while discarding others, solely because they have a child of one sex and want one of another. The Society demands that fertility clinics abide by the decision and most have indicated that they would ("Fertility society choosing embryos just for sex selection." *New York Times*. 16 February 2002). However, other techniques of sex selection, for example, sperm sorting," are not affected by the ruling of the Society.

23. The now widely debated issue of genetically modified (GM) crops and GM food is representative of the kind of conflicts that are bound to erupt more widely in struggles between interested organizations and groups that want to extent the use of transgenic crops and consumers for example that desire to not only closely monitor its use but an outright ban of such crops and foodstuff that incorporates genetically modified crops. Despite the controversies and health or environmental concerns surrounding the use of GM crops and the eroding trust in GM foods among consumers (although there are marked differences in the consumer response from country to

country), the estimated global area of transgenic crops continues to grow according to an annual survey of the commercialized GM crops carried out by Clive James, Director of the International Service for the Acquisition of Agri-biotech Applications. The estimated global area of transgenic crops for 2001 is 130 million acres grown by 5.5 million farmers (up from 3.5 million in 2000) in thirteen countries on all continents. In 2001, the United States grew the largest portion, sixty-eight percent of the total worldwide, followed by Argentina and Canada (mainly soybean, com, canola, maize). The increase between 2000 and 2001 is 19 percent, equivalent to 20.8 million acres. Between 1966 and 2001, the global area increased more than thirty-fold (James 2001).

24. I have examined these changes of and consequences for modern societies in greater detail in *The Fragility of Modern Societies* (Stehr 2001).

References

Baldi, Pierre (2001) *The Shattered Self: The end of natural evolution*. Cambridge: MIT Press.

Bell, Daniel (1960) *The End of Ideology* Glencoe, III.: Free Press.

—— (1973) *The Coming of Post-Industrial Society: A venture in social forecasting*. New York: Basic Books.

Böhme, Gernot (1992) *Coping with Science*. Boulder, Colo.: Westview Press.

Brooks, Harvey (1965) "Scientific concepts and cultural change," *Daedalus* 94: 66–83.

Cerny, Philip G. (1991) "The limits of deregulation: Transnational interpenetration and policy change," *European Journal of Political Research* 19(2,3): 173–96.

Chargaff, Erwin (1975) "Profitable wonders. A few thoughts on nucleic acid research," *The Sciences* 17: 21–6.

Dasgupta, Partha S., and Paul A. David (1986) "Information Disclosure and the Economics of Science and Technology," in *Arrow and the Ascent of Modern Economic Theory*, George F. Feiwel (ed.). London: Macmillan. 519–42.

Fukuyama, Francis (2002) Our Postmodern Future. Consequences of the biotechnology revolution. New York: Farrar Straus and Giroux

Fukuyama, Francis, and Caroline S. Wagner (2000) *Information and Biological Revolutions: Global governance challenges—Summary of a study group*. Santa Monica, Cal.: Rand.

Gaisford. James D., Jill E. Hobbs, William A. Kerr. Nicholas Perdikis, and Marni D. Plunkett (2001) *The Economics of Biotechnology*. Cheltenham: Elgar.

Galbraith. John K. (1983) *The Anatomy of Power*. Boston: Houghton Mifflin.

Green, Harold P (1976) "Law and Genetic Control: Public-policy questions," in *Ethical and Scientific Issues Posed by Human Uses of Molecular Genetics*, Marc Lappé and Robert S Morrison (eds.). Annuals of the New Academy of Sciences 265. New York: New York Academy of Sciences.

Habermas. Jürgen (2001) *Die Zukunft der menschlichen Natur: Auf dem Weg zu einer liberalen Eugenik?* Frankfurt am Main: Suhrkamp.

Hindmarsh, Richard, Geoffrey Lawrence, and Janet Norton (1998) "Bio-Utopia: The way forward," in *Altered Genes: Reconstructing nature—The Debate*,

Richard Hindmarsh, Geoffrey Lawrence, and Janet Norton (eds.). London: Allen and Unwin. 3–23.

Horowitz, Irving L. (1985) "Elite roles and democratic sentiments," *Society* 22: 16–19.

James, Clive (2001) "Global review of commercialized transgenic crops: 2001," *ISAAA Briefs* (24).

King, Lauriston R., and Philip H. Melanson (1972) "Knowledge and politics: Some experiences from the 1960s," *Public Policy* 20: 83–101.

Lane, Robert E. (1966) "The decline of politics and ideology in a knowledgeable society." *American Sociological Review* 31: 649–62.

Lanza, Robert P., Arthur L. Caplan, Lee M. Silver, Jose B. Cibelli, Michael D. West, and Ronald M. Green (2000) "The ethical validity of using nuclear transfer in human transplantation," *Journal of the American Medical Association* 284: 3175–9.

Lowe, Adolph (1971) "Is present-day higher learning 'relevant'?," *Social Research* 38: 563–80.

Lübbe, Hermann (1977) *Wissenschaftspolitik: Planting, Politisieren, Relevant.* Zurich: Interfrom.

Luhmann, Niklas ([1991] 1993) *Risk: A sociological theory.* New York: de Gruyter.

Mulkay, Michael (1993) "Rhetorics of hope and fear in the great embryo debate," *Social Studies of Science* 23: 721–42.

National Bioethics Advisory Commission (1999) *Ethical Issues in Human Stem Cell Research.* Washington, D.C.: National Bioethics Advisory Commission.

Pavitt, Keith (1987) "The objectives of technology policy," *Science and Public Policy* 14: 182–8.

Rammert, Werner (2000) "Ritardando and Accelerando in Reflexive Innovation, or How Networks Synchronise the Tempi of Technological Innovation," Working Paper TUTS-WP-7–2000. Technical University Berlin.

Sinsheimer, Robert L. (1976) Recombinant DNA—on our own," *BioScience* 26: 599.

—— (1978) "The presumptions of science," *Daedalus* 107: 23–35.

Stehr, Nico (2001) *The Fragility of Modern Societies: Knowledge and risk in the information age.* London: Sage Publications.

Vanderburg, William H. (2000) *The Labyrinth of Technology.* Toronto, Ont.: University of Toronto Press.

Wright, Susan (1986) "Recombinant DNA technology and its social transformation," *Osiris* 2: 303–60.

Zuckerman, Harriet (1986) "Uses and control of knowledge: Implications for the fabric of society," in *The Social Fabric: Dimensions and issues*, James F. Short Jr. (ed.). Beverly Hills, Cal.: Sage Publications 334–48.

Part I

The Emergence of Knowledge Politics: Origins, Context and Consequences

Introduction

Reiner Grundmann

Gernot Böhme starts this section with the observation that current developments in modern medicine, especially reproductive medicine, transplantation medicine, and genetic technology, have turned the human body into the object of theoretically limitless possibilities for manipulation. For Böhme, the experience of nuclear energy sets the example that there should be regulation of knowledge. Taking nuclear energy as the template, he applies this to the case of manipulating human nature. While he thinks that the necessity of regulating the generation and the application of knowledge in this context is incontestable, he rejects conventional ways of dealing with this. Conventional policies would distinguish between science and society and conceive of regulation as coming from outside science, that is, through laypersons and politicians Böhme holds that such a regulation from outside is futile, as it comes too late, both in a temporal sense (the application of knowledge, once produced cannot be restrained in a global market place), and in a logical sense, as it comes after the act of knowledge creation by scientists. Hence, the question arises as to whether the regulation of knowledge might not be a task for scientists themselves, and whether the presence of ethical points of view ought not already to be implicit in their day-to-day activities.

First, Böhme alerts us to the fact that in premodern times there was no need for ethical controls of our dealings with nature, since this relation was culturally embedded. From classical antiquity into the Middle Ages, agriculture and mining, for example, were embedded in religious rites "In antiquity it was individual gods, with particular areas of competence, with whom one had to make contact; in the Christian context there was an order to creation, which had to be respected. Violations of the reigning order had to be atoned for; witness such figures as the chained Prometheus or Jonah in the whale." But how

3

can we learn from these practices, especially as there seems to be no way back in history?

Böhme addresses the question of how a regulation of science could be envisaged from within science, by developing a form of knowledge that already implies a moral respect for nature. Which form could this take? Böhme contends that empathy is the key notion here. He refers to Hans Jonas, who developed an analogy between natural beings and infants: "The newborn, whose mere breathing incontrovertibly lays an obligation on its environment, namely: to care for it. Just look at it and you know this." The analogy implies a reintroduction of teleology into our understanding of and dealings with nature. This means to acknowledge intention in organisms, "the intention to continue living and to develop. Organisms are conceived of as entities that take an interest in their existence. And when we conceive of them as such, then we have no choice but to take this interest into account ethically." Böhme refers to Herder, who in his critique of Kant developed the concept of *knowledge as acknowledgment*, in knowing nature, we realize it is of our kind, and thereby acknowledged. In knowing nature, one's own naturalness is at the same time acknowledged. And Böhme seeks support from another German classical author. Goethe, who pointed out that "living beings, being what they are, are never in a single moment completely in existence; they are, as it were, smudged over time. They have, to draw an analogy to a different temporal entity, the character of a melody. A melody too, after all, cannot be represented in the moment by that which it is at the time, namely by the single note. To understand the melody, one must follow it through time." Goethe uses the term *Bildung* to describe the fully lived lifespan "Thus, too, the man of science has at all times been distinguished by a drive to recognize living developments [*Bildungen*] as such, to grasp their outwardly visible, tangible parts in their interconnection, to accept them as intimations of their inner workings and so to master the whole, as it were, by means of contemplation."

Drawing upon the German Romantic tradition, Böhme shows that "as long as we maintain that what nature *is* is determined exclusively by modern science, then we will be dealing with a nature that is not ethically relevant for us." However, he is also aware of the severe limitations of the practical efficacy of such a new ethics of nature; "Against the possibility of doing things differently there stands the actual power of the established business of science and in particular its integration into strategies of economic exploitation. Even were many bioscientists convinced of the possibility of an ethically relevant knowledge of nature, the question still arises: With what material forces could their ideas be combined, in order to have a real chance of being accepted? This would probably only be possible if this other knowledge of nature could make *curative promises* comparable to the established life sciences." Böhme thinks not all is lost—he *conceives of a mobilization from within*, that is, a forceful movement among the scientists themselves. This could be triggered by growing public pressure.

Van den Daele's contribution offers a counter position to Böhme's. He introduces Polanyi's concept of personal knowledge, which he primarily sees as a corrective to the one-sided picture of knowledge as impersonal, disembedded, and information-like. "Even in science, digesting the information communicated in a publication does not necessarily mean that one *knows* what is published—if knowing implies that one knows how to do the science, that is, reproduce the experiments reported, use the results, etc. To that end, theories, data, explanations, algorithms, rules written on paper, or disseminated through the Internet are not enough. Personal knowledge is necessary, this must be acquired through socialization and a local context or culture that is conducive to the kind of knowledge that is published (instruments, orientations, standards). Strictly speaking, knowledge is universal only to the extent that the personal competencies and the local circumstances that produce (or reproduce) that knowledge are universal."

But then he goes on to affirm that "the growth of modern science has in many fields replaced the implicit and embodied knowledge of the practitioner by the explicit and disembodied knowledge of the scientist." This argument provides him a platform from which to counter claims about the inherent connection between values scientists may hold and their scientific work. For him, modern scientific and technical knowledge is information and "as such culturally disembedded, which means: differentiated from encompassing moral values, social concerns, and political goals." This is in rather stark contrast to Böhme, who thinks such a link is not only possible but necessary.

As van den Daele explains, once ethics and knowledge have become separated, there is no way of regaining control from within science. If there are to be any restrictions, they will come from outside science in the form of legal duties, economic interest, or professional codes (although it would be interesting to see how professional codes of scientists could be understood as an example for a combination of knowledge and ethics).

However, according to van den Daele, it would be wrong to assume that we have lost the deeper meaning of nature and the perception of the sacred in things. It is only that such knowledge is dissociated from and external to the instrumental knowledge of nature accumulated through science and technology, it exists alongside it, so to speak, but is left behind by developments in science. Van den Daele rejects the idea of an alternative natural science. For him, this remains a dream that has never come true. Consider the case of ecology: it did not turn out to be a subversive science, nor an alternative to the Baconian vision of nature as a resource and of natural knowledge as instrumental power. It is true that ecology sees nature as an evolving, interconnected self-regulating system that is different from the clockwork universe of classical physics. However, this does not mean that we have moved to a non-instrumental, contemplative, or moral view of nature. We have rather moved to a more sophisticated technological view. Ironically, ecosystem approaches will help make Bacon's vision become more real than before, in that nature can only be commanded by obeying its laws. Therefore, "scientific ecology does not transcend the Baconian vision of the domination of nature, it perfects it."

In the third paper in this section, Steve Fuller addresses die problem of governance of knowledge from another angle. As he makes clear at the outset, "It does not make sense to speak of the governance of knowledge without specifying knowledge-bearing institutions." Usually such an institutional perspective is absent from analysis. People seem to think that the same knowledge may emerge from very different institutional settings: universities, business firms, or government laboratories. Fuller thinks this is a cheap solution to the society-knowledge problem. When addressing debates over whether knowledge producers should govern themselves or be subject to external control. Fuller examines Böhme and van den Daele's contributions, two scholars once known for their involvement with the German finalization movement.

Fuller's verdict on Böhme is that his outline of a new ethics of nature leaves open the question of institutionalization. Fuller would hardly content himself with simply encouraging mature scientists to delve into nineteenth century *Naturphilosophie*. He sees two possibilities, either making science students study the histories of their fields, so that they learn of alternative frameworks for casting their inquiries before they have been indoctrinated into the dominant framework, or involving a broader range of scientific practitioners into the governance of their own fields (the more scientists see themselves as distant from cutting-edge specialist research, the more open they are to unorthodox concerns). However, Fuller thinks that both possibilities would hardly be practicable, as these would run counter to the scientists' self-interest (challenge to their professional identities or specialization).

Hence he looks at van den Daele's proposal that science's applications be subject to external control, especially so as not to interfere with the promotion of alternative, so-called indigenous knowledges. Fuller thinks that such a proposal presupposes that clear distinctions can be drawn between science and other forms of knowledge, as well as between research and its applications. It is obvious that he is skeptical about making such distinctions. Furthermore, he does not think that van den Daele's vision of a state-enforced cultural pluralism offers an outline of institutional design that might realize it.

Fuller focuses on the university as an institution in order to resolve some of the above mentioned problems. Drawing on Schumpeter's phrase of creative destruction, he sees universities as very much engaged in an endless cycle of creating and destroying "social capital" by virtue of their dual role as producers and distributors of knowledge: "As researchers, academics create social capital because intellectual innovation necessarily begins life as an elite product available only to those on the cutting edge. However, as teachers, academics destroy social capital by making the innovation publicly available, thereby diminishing whatever advantage was originally afforded to those on the cutting edge."

Taking up Fred Hirsch's notion of "positional goods" and Werner Sombart's analysis of symbolic consumption as a status symbol, Fuller claims that "an expanded production of positional goods, combined with increased efficiency in the production of material goods, results in systemically irrational outcomes that we have come to expect (and perhaps even rationalize) as our 'knowledge society.' Specifically, the resources spent on acquiring credentials and marketing goods come to *exceed* what is spent on the actual work that these activities are meant to enhance, facilitate, and communicate." This could mean that more and more people, in order to enhance their status, are drawn to study at university, at the same time lowering the value of the degree (as more people have them). The same logic applies to universities competing for kitemarks (research and teaching audits and the resulting rankings).

Applying this model to science-based innovations, Fuller observes that "the main economic impact of a successful invention is that it destabilizes, or creatively destroys, markets, as more people seek patents, everyone else will soon have reason to engage in the same activity in order to restore their place in the market. Thus, a lethargic economy dominated by rent-seekers is quickly transformed into a dynamic commercial environment." It takes only one more step to reapply this model again to the university.

1

Knowledge Policy as the Task of Science: On Ethically Relevant Knowledge of Nature

Gernot Böhme

The tremendous increase in knowledge produced in the last few years, above all in the life sciences, has not only furnished humanity with new possibilities for acting, but also given rise to a fear of these possibilities (Böhme 200la). The experiences that humanity has gained from our power to command nature's energies by technological means—here I refer to our experiences with nuclear energy—have taught us that there must be something like knowledge policy and the regulation of knowledge in order to frame these actions in a reasonable manner. This is much more the case now that life itself, and with it our own individual lives as well, is becoming easier to manipulate on the basis of knowledge gained from science. The necessity of regulating the generation and the application of knowledge in this context is incontestable. The question is only at what point, and through whom, this regulation should occur. In answering this question, a well-known and apparently taken-for-granted dichotomy proves to be an obstacle; namely, that between scientists and laypersons.

Scientists have a natural, and moreover a constitutionally guaranteed, claim to freedom of research. They themselves, as a result, are only driven to so-called self-restrictions by means of external pressure. Laypersons and their democratic representatives have an entirely genuine interest in the regulation of science, which, however, relates in the end not to the content of knowledge, but rather only to its application. The crucial problem that hinders an effective regulation of knowledge in the end is the fact that such regulation from outside practically always comes too late—and that in a double sense. On one hand, the application of knowledge, once produced, cannot be hindered in a globally pluralistic and market-driven world society. On the other hand, any regulation, whether it is motivated by utilitarianism, religion, or even considerations of human rights, remains external to knowledge itself. A legally and ethically neutral access to the object of scientific inquiry, in particular to life, cannot be morally constrained by means of subsequent regulations. Thus, in the case of so sensitive an object as life, especially human life, the question arises whether the regulation of knowledge might not be a task for scientists themselves, and whether the presence of ethical points of view ought not to be implicit in their approach to their object. It is to this question that the following article is dedicated.

Ethics of Nature

Our dealings with nature require an ethical regulation—there is something astonishing about this sentence; or there would be, if astonishment had not long since been washed away by a flood of ethical approaches on offer. Bioethics, ecological ethics, animal ethics, medical ethics compete to remedy this deficit. Ethics committees are supposed to obtain legitimization, whole institutes to produce ethical knowledge, and new professorships to disseminate this knowledge. At the same time, there is a great deal of embarrassment when it comes to describe just what an ethics of nature should look like. Some attempt to subsume it under an ethics of the good life: Unspoiled nature is simply a component of the good life, and without it life would no longer be quite so good (e.g., Seel 1991). Others attempt to bring utilitarian ethics to bear, sensibly trying to weigh up the benefits and costs to individuals, and to the masses, of one manner of dealing with nature or the other. Yet others back an ethics of compassion, and want to extend the scope of entities deserving of ethical consideration to all beings capable of suffering. The most radical are probably those who, in a contract with nature, would grant rights to the animals as well, and over and above that might extend this legalistic approach to nature even to minerals (Meyer-Abich 1997). There is no consensus among the moralists, and therefore we might do well to return once more to the astonishment that an ethical regulation of our dealings with nature is even necessary.

This astonishment can be articulated in two ways; two ways, however, that are diametrically opposed. If I call these two ways to articulate this astonishment "premodern" and "modern," it might be thought that only the latter could possibly be of interest to us moderns. This, of course, is the way one thinks if one conceives of history as a progressive, linear process. I do not do this; on the contrary. I believe that even in our own time, premodern modes of thought are present and significant. In premodern times there was no need for the ethical regulation of our treatment of nature, because this treatment *was* regulated. In premodern times (and I mean this now in the historical sense; from classical antiquity until well into the Middle Ages), agriculture and mining, for example, were bound up in religious rites that articulated human beings' dependence on nature and enveloped their behavior with prayer, incantation, gratitude, and obligation, kept it within bounds and protected it from hubris. In antiquity it was individual gods, with particular areas of competence, with whom one had to make contact; in the Christian context there was an order to creation, which had to be respected. Violations of the reigning order had to be atoned for; witness such figures as the chained Prometheus or Jonah in the whale.

The modern astonishment at the fact that our dealings with nature demand to be ethically regulated stands out against a background of completely opposing commonplaces. It is not the commonplace of an ethical view of nature that constitutes this background, but rather the idea that an ethical relationship with nature is pointless or even absurd. This astonishment can be no better illustrated than by the well-known anecdote about the Cartesian philosopher Malebranche. When a visitor begged Malebranche s pardon because he had inadvertently stepped on his host's dog, who gave a howl, Malebranche replied. "What do you mean, monsieur, animals have no soul." To the modern mind natural beings are pure objects—in Cartesian terms, extended substances—and, as such, not the object or target of ethical reflections. Ethics, on the other hand, is understood as something explicitly directed only at persons, which has as its function the regulation of relations between human beings. Both positions, namely that of a religiously interpreted dependence on nature on the one hand and that of an objectivist conception of nature on the other, combine to shape current approaches to an ethics of nature. As a result, these approaches appear in part fundamentalist and esoteric and in part utilitarian. That means, however, that such ethical endeavors attempt to bring us back to old commonplaces, whether of a premodern or a modern fashion. And yet this will not be so easy, in the first place due to the incompatibility of these solutions, and still more because we have encountered difficulties with both the mythological-Christian conception of nature and the scientific-technical conception of nature. These problems are not only of a practical type that could be solved practically, if need be: rather, they face us with serious questions, and serious questions are moral questions (see Böhme 2001b).

Serious questions, in one's individual life, are those questions that, by one's answers to them, contribute to determining what kind of person one is, or how one is as a person. We are faced with such questions nowadays in terms of our own nature, that is, the body as that expression of nature that we ourselves are. The development of scientific-technical medicine, especially reproductive medicine, transplantation medicine and genetic technology, has now turned the human body into the object of theoretically limitless possibilities for manipulation. Thus, in principle, what it is about human beings that might still be described as nature is called into question. For the individual this becomes a moral question, inasmuch as each of us, at some time in our lives, will have to decide to what point we want to implement medical-technological manipulation: or, inversely, what of ourselves we are willing to accept as nature, that is, as given. Within the framework of this debate, such moral topoi are named as death with dignity; natality as a component of human dignity; the right to ignorance; the right to imperfection; the right to fortuitousness; and the entitlement of every individual to be, of himself, a radically new beginning. These debates can presuppose no firm concept of human dignity; on the contrary, only in the face of the challenge of medical technological progress can what is meant by human dignity be articulated. The general result of these debates may be that naturalness, as an essential component of being human, has only become clear through the provocation of modern medical and biotechnological possibilities: today, the dignity of the human being is being threatened above all in terms of his body.

So much for the serious questions that concern the individual, serious questions are posed for society whenever the act of decision at the same time determines in what kind of society we live, and what basic assumptions we share regarding human beings and society. Such questions are posed for us today—not only, as for every individual, in relation to that nature that we ourselves are, but also in relation to nature external to us. Here I refer to our treatment of animals, of the landscape, of ecotopes, and natural resources. That this is not merely a matter of expediency and a pragmatic need for regulation, but rather of moral questions, has been demonstrated by the debates surrounding the law for the protection of animals and surrounding nature as a constitutionally protected good. The new law for the protection of animals in the Federal Republic of Germany demands the protection of animals as "fellow creatures." This formulation seems to contain, above all, a new respect for animals and a rise in their status. Yet, what is more important, the creature-nature of human beings themselves, as the chief basis of their relationship to other natural creatures, is being articulated. The introduction of nature conservation into the German Basic Law was a serious and thus a moral question. Indeed, it affects our conception of the state: conservation as protection of the *natural bases of life* is introduced as a new function of the state, in addition to its previous functions. These were: the guarantee of external and internal security (security state), the organization and guarantee of the justice system (state founded on the rule of law), the organization and fostering of the education system (cultural state), the social ensuring of the citizens' safety (welfare state), and finally the direction of the economy (social market economy). For the first time, by introducing nature conservation as a function of the state, an understanding of our state- organized life is articulated that expresses that we live in and on the basis of nature (Böhme 2001b: 155–31).

Ethics and Cognition

The relative progress that has been made toward taking nature seriously in moral terms, both our own nature and nature outside ourselves, should not obscure that an ethical point of view toward nature remains relatively superficial. The way in which we conceive of and recognize nature obviously offers us no grounds to relate to nature in a moral fashion. What nature *is* is still conceived as ethically neutral, and moral attitudes toward nature obviously stem from other sources. The significance of these other sources, whether they be creation theology or the rediscovery of the body as that nature that we ourselves are, tentatively brings new points of view into our conception of nature—points of view, however, that have not yet been developed into forms of knowledge. The only relevant knowledge of nature is constituted, as ever, by science. And it offers no occasion whatsoever to conceive of its objects as ethically relevant.

A characteristic example of this dilemma is the hair-splitting with which many vegetarians argue just how far one should carry vegetarianism (Wolf 1990; cf. Böhme 1991). Meat may not be eaten; fish can be, however, because they lack a central nervous system. Here *objective* reasons for a natural creature's capacity for suffering are given, while the empirical context in which one can encounter a fish as a living being remains completely unaccounted for A fish can appear to one, in its liveliness, in a manner that makes it difficult for one to kill it—and what one knows scientifically about fish, in such a situation, can be only a sophistic legitimation for nonetheless killing it.

This example demonstrates that the moral attitude toward a natural being can certainly be better developed than the knowledge of this being. It raises the question whether there is a knowledge of nature in which respect for natural beings also determines the ways and means of gaining knowledge. This question also concerns the three examples mentioned, in which taking nature seriously in our society is already to some degree a matter of consensus. The insight, gained from the problem of the environment and from technological access to the human body, that we ourselves *are* nature finds no basis as yet in any knowledge of this nature of ours, namely, the body. The body *qua* physical body is conceived in terms of anatomy, physiology, neurophysiology, genetic technology; and such a concept of our own nature offers no grounds whatsoever to treat even this nature of ours with respect. Why should one respect one's own heart, if it is nothing but a pump that can be replaced? Where is respect for animals to originate, if a science pictures them as neurophysiological machines and correspondingly uses them in experiments? And where are we to find respect for nature external to us, if it is seen only as the basis of human life?

This point needs to be explained more precisely. In what sense is respect[1] a prerequisite for a moral attitude toward natural creatures? And in what sense is our scientific knowledge lacking in respect for natural creatures? To deal with the last question first; It is certainly possible to say that modern science has by no means undermined, but indeed rather increased, the respect that we show toward nature. In fact, what we have discovered in terms of interdependencies in nature becomes ever more astonishing as science progresses. Naturally, every step further terminates our astonishment at the step before, explaining the functioning of that which seemed to us at first incomprehensible. To be sure, this has not meant that astonishment in the face of nature has retreated along the fronts of research. The total impression of natural occurrences, such as evolution or cell metabolism or ontogenesis of the individual creature, for instance, remains something that compels our respect. It has nothing to do with respect in the moral sense, of course, but rather with astonishment and admiration. In any case, no researcher, on the basis of his insights into these interdependencies, is going to develop any reluctance to intervene in them, alter them, and develop them further. On the contrary: the very type of knowledge he has (his recognition of the life processes as functional interdependencies) is also intended to intervene in them or to alter them. He gains his knowledge of these interdependencies in no other way than by means of intervening in them and altering them. For this reason, it is quite rightfully said that biochemical and biogenetic research, as such, are already technology. In pursuing them, one and the same researcher can perfectly well cultivate yet another, quite different, attitude toward nature. It is well-known, for example, regarding experiments on higher animals, that workers in the appropriate research centers care for them lovingly, even give them names; and yet nonetheless implant electronic probes in their brains and stimulate their sexual behaviors by means of transmitters. The behavior patterns of human beings need not be thoroughly consistent with one another. Loving behavior towards the experimental animals does not correspond, in any case, to the scientific knowledge that is gained from them.

Why, however, is respect for natural creatures the basis of moral conduct toward them? Or is it that at all? Moral conduct can certainly come into being on the basis of principles. Thus, one can demand, based on insight into one's own naturalness, that animals be treated as fellow creatures. And that might very well be sufficient, if it is a matter of canonizing a social consensus, that is, of prescribing laws. This does not mean, however, that animals will be treated with respect in particular cases: rather, a law such as the law for the protection of animals may become no more than the bureaucratic constraint on a basically unchanged attitude toward animals. That is to say: in order that we refrain from a certain behavior toward animals not simply because it is forbidden, but rather because we shrink from behaving toward animals in such a manner, animals themselves must be seen in a different light.

The same is also true for that nature that we ourselves comprise: the human body. The general view that nowadays, through progress in biomedical technology, the dignity of the human being is threatened in terms of his body, does not at all mean that we consider ourselves, insofar as we are nature, with respect. Self-respect is still directed at the self *qua* person, not at the self *qua* body. It is necessary to practice the latter, however, if the general and abstract view that naturalness is an essential part of our being human is also to take on practical reality. Now, as a rule, we know our bodies not in the way that we ourselves are them, but rather exactly in the way that we are not them, namely as an object of science, medicine, or pharmacology. And it is also from this perspective that we behave toward ourselves *qua* nature. What is once again lacking here is a knowledge, in this case a knowledge of the human body, that implies respect for it, because it has already been gained under the auspices of self-respect *qua* body. What form should such a knowledge take? This type of knowledge contains, over and above the difficulties inherent in an ethically relevant knowledge of nature, the additional problem that it ought to be a knowledge of ourselves, insofar as we are not in control of ourselves. If we wish to understand our embodiment as an experience of nature, then obviously the acknowledgment of the unavailability of this nature must be contained within it.

The task we must set ourselves, then, does not consist of providing an ethical superstructure for a nature still perceived in terms of modern science: rather, it consists of developing a form of knowledge that already implies a moral respect for nature. How might such a knowledge look?

What Is and What Ought to Be: Hans Jonas

By no means is such a form of knowledge unknown. The above mentioned example of Malebranche's dog makes this quite clear. From hearing the dog howl, we know very well that it is suffering. The idea of the howling dog is inherently a compassionate idea, and demands a certain behavior of us. Such knowledge can quite easily be glossed over and displaced, as indeed the historical example demonstrates; and in particular it is not sufficiently developed to be able to compete, if necessary, with the positivistic knowledge of science. What is worse; its potential is theoretically denied.

The usual argument against such a form of knowledge is the dogma that from what is, what ought to be cannot be deduced Hans Jonas' demonstrated that this dogma conceals a circular argument. Being is conceived therein, namely, in positivistic terms: as an unrelated and self-sufficient fact. Contrary to this tendency, Jonas highlights that among oilier things, we are dealing with a form of being that as such, in that which it is, already makes claims on us. For him, the prototype for this mode of being is the human infant: "The newborn, whose mere breathing incontrovertibly lays an obligation on its environment, namely: to care for it. Just look at it and you know this" (Jonas 1988: 235).

Jonas no doubt knows very well that one can also see an infant differently, but does one then see it as an infant?

> What really and objectively is *there* is a conglomeration of cells, which are in turn a conglomeration of molecules with their physico-chemical

transactions, which can be recognized as such together with the *conditions of their continuation*; but the fact that this continuation *should be*, and therefore that someone should do something to ensure this, is not part of the result and cannot in any way be *seen* from it. Definitely not. But is it the infant that is being seen here? (ibid. 236)

One can see from this quotation that Jonas is making an effort to take knowledge of the living world seriously. He does this in Heideggerian fashion, by differentiating among modes of being and the modes of knowledge attributed to them. An infant is, in that which it is, not merely something that exists, but rather a something that in its very existence expresses a claim on its environment. Of course, it cannot be overlooked that this way of seeing demands that we also see ourselves, that is, the subjects of knowledge, differently. In practical terms, it is assumed that we, these subjects, are open to the appeal that emanates from a being of the type represented by the infant; or to be precise, a type of engagement in the world in which an infant is already understood in principle as belonging to us. To say it briefly: an understanding of subjects as human beings—and this, as is made quite clear by Kant, for example, is just how the subjects of objective knowledge are not understood.

Jonas outlines—proceeding from the conquest of the dogma that from what is, what ought to be cannot be deduced—a new philosophy of nature that in principle allows the function of appeal to be attributed to non-human beings as well. This is chiefly a matter of a rehabilitation of teleology in the knowledge of nature; that is, a liberation from the as-if-status into which Kant banished it. Since Kant, indeed, the concept of expediency in the life sciences has only a methodological status (Kant 1790). That is: it is indeed used as an heuristic and as a view that can develop material, but it is not understood as a characterization of objects, and therefore, of the organisms themselves. This has led to a general reluctance to use the terms "teleology" or expediency, and instead rather to speak of teleonomy, function, adaptation, etc. These are newer versions of Kant's statement that the concept of expediency is not a concept that is constitutive for the object, in this case the natural being, but rather that it merely represents a regulative concept for the continuation of research. As honorable as this attitude is from the methodological point of view, it is just as awkward with regard to the object, in particular the organisms. For this methodology uses a knowledge of organisms obtained in the way living beings are approached in the lifeworld, without acknowledging this knowledge as knowledge.

It is thus no wonder that under the present stress of the problems into which we have fallen using the objectivist approach to nature, there have been efforts to rehabilitate a teleological view of nature. Here I wish to draw attention to the work of Spaemann and Löw in particular (1991). Jonas' work moved in this direction as well. His particular achievement, however, is that from the outset he saw the teleological view of nature in the context of an ethics of responsibility. For him it was a matter of finding an approach to natural beings, particularly organisms, analogous to our relationship to infants. The teleological view means, for Jonas, a method of approach to organisms in which an intention is acknowledged in the organisms: the intention to continue living and to develop. Organisms are conceived of as entities that take an interest in their existence. And when we conceive of them as such, then we have no choice but to take this interest into account ethically.

From the work of Jonas there also arise connections to other important efforts to conceive of natural beings, particularly organisms, differently from the way they are conceived in modern science. The first of these that I would like to mention is Heidegger's *descending ontology* of organic matter. Already in the 1930s, Heidegger was the first to suggest developing the ontology of nature not as an ascent from the elementary to the complex, but rather the reverse: as a descent from the complex to the elementary (Heidegger 1957: esp. sec. 10; 1983 [1929/30]). According to this suggestion, what nature is would be most apparent in human beings: and from their ontology—that is, from what for Heidegger is the fundamental ontology—one ought to develop the strata of organic matter subtractively, as it were, down through the mineral world to the basic elements. On this view, in the sense of Heideggerian ontology, the lack of a particular potential for being is not merely its pure non-existence; rather, it is frequently operative in the form of a deficient method. Most importantly for our context, the structure of care of *Dose in,* that the human being in his being concerns himself with his very being is also to be found in modified form among the organisms. In this Jonas follows Heidegger, who could also quite easily be regarded, concerning the descending ontology of organic matter, as his student.

To emphasize this point: descending ontology is a sort of anthropic principle.[2] What nature is in general is judged by that one of its products that is, in a certain sense, late in evolutionary terms and displays the most complex structure: human beings. From this premise organisms are at least believed capable of the structure of care developed in human beings. Organisms are to be regarded such that in conceiving of them, their interest in their own existence is acknowledged.

The concept of acknowledgment is evidently a conception of knowledge in which the relationship to ethics is already implicit. I would therefore like to take up in greater detail the design of an alternative way of knowledge— alternative to modern natural science, in which the concept of acknowledgment plays a central role. I refer to the epistemology of Johann Gottlieb Herder.

Knowledge as Acknowledgment: Herder

Kant, in his *Critique, of Pure Reason* and the *Metaphysical Foundations of Natural Science,* presents the hitherto most convincing support for science in the modern sense (see Böhme 2002). Herder's critique of the Kantian theory can thus correspondingly be regarded as one of the most important contributions to the critique of the type of knowledge represented by modern science. It is to be found in Herder's book *Reason and Experience: A Metacritique of the Critique of Pure Reason* (1799). Kant describes knowledge essentially as a process of appropriation: We appropriate to ourselves, in the form of knowledge, that which is of itself alien to us. This conception of knowledge culminates in the provocative statement: "We prescribe the laws to nature." We are given no more than an incoherent manifold and it is we ourselves who bring order and coherence to nature. In contrast, Herder maintains that knowledge is a *becoming aware of the unity in multiplicity (Innewerden des Einen im Vtelen;* ibid.: 196). To quote Herder:

> You can know nothing where there is nothing to know; you can bring nothing together within you where there exists nothing brought together by nature. The function of reason [*Verstand*] is: to acknowledge what is there, to the extent that it is understandable to you, that is, that it appertains to your reason. (p. 208)

In this quotation, the first important element of Herder's epistemology appears: He regards unity as given by nature. This, indeed, was the dimension in which Kantian epistemology, as Kant more closely approached concrete nature in the *Critique of Judgment,* got into difficulties: Nature organizes itself by itself into unities. In the face of these given unities, namely those of organic beings, Kant could only react with the methodology of "as-if." As the second important element in this quotation from Herder, there already appears the important term *acknowledgment.*

With the connection between knowledge and acknowledgment Herder, as a linguist, refers back to an antiquated usage of *knowledge (Erkennen)* as found, for example, in the biblical, that is. Lutheran formulation *Adam knew (erkannte) his wife.* To know someone, in this sense, means to acknowledge him in his belonging to oneself I quote Herder on this etymological connection:

> Instead of free-roaming discursive thinking, let us then rather say to know [*Erkennen*], and that in the original power of the word. To know, to recognize, to acknowledge [*Kennen, erkennen, anerkennen*] is a term of kindredness; it means to feel, inwardly and quickly, a commonality of nature and species, phylum and descent. Thus I know [*erkenne*] my child, a friend, my beloved; thus I know truth, good, beauty. I acknowledge them [*Ich erkenne sie mir an*], and say loudly: they are of my nature, of my kind, whereby then all spontaneous seeming and semblance ceases of itself. (p. 205)

By means of this quotation, a further important factor in Herder's concept of knowledge as acknowledgment becomes clear: Where, in the statement against Kant, it first became apparent that unity is acknowledged not as an achievement of reason, but rather as given, so now we see that acknowledgment means something not only for the given, but also for him to whom something is given, that is, of the subject. According to Herder, in the cognitive process one's own nature is not projected into the object of knowledge: rather, it is rediscovered within the object. The object of knowledge is understood only to the extent that it is related to me. This means for nature: In knowing nature, one's own naturalness is at the same time acknowledged. We have already touched on this point repeatedly: We will not be able to manage a change in our attitude toward nature if it is not accompanied by a change in attitude toward ourselves. The acknowledgment of nature in its own being will not succeed if we do not at the same time carry out the acknowledgment of our own naturalness.

This statement of Herder's, to the effect that knowledge of nature *qua* acknowledgment is simultaneously the acknowledgment of one's own naturalness, opens up important possibilities for epistemology. The first is the possibility of overcoming the reproach of anthropomorphism. The possibility of an ethics of nature or, indeed, even of an ethical view of nature, is continually countered by the argument that nature is thereby seen in anthropomorphic terms. According to Herder, the transference onto nature of concepts we have developed through the experience of ourselves does not constitute anthropomorphism, because these concepts actually spring from our acknowledgment of ourselves as nature. He demonstrates this with the concept of force (*Kraft*). Since the concept of force entered into natural philosophy or the natural sciences, of course, from Leonardo through Kepler to Newton, there have been attempts to expel this concept once again as an anthropomorphism. Herder now explicitly concedes that this concept originates from self-experience: "A combination of force and effect constitutes our feeling" (p. 230 f.). But this feeling is in fact simply an expression of our self-experience as natural beings. It would therefore be a denial of ourselves if we wished to renounce this concept in our experience of nature: "It [reason] would therefore have to forget and destroy itself, if it could for even a moment disregard the law through which it exists: *Cause creates effect*" (p. 231).

At this point there follows the second important possibility for a knowledge of nature. In the knowledge of nature—explicitly, nature external to us—we can proceed from the fact that we experience ourselves as natural beings. What nature is, we experience in ourselves and from this we can therefore also give evidence of what it means for other natural beings, particularly for organisms, to exist. At this point it is especially important to emphasize that we are referring here to a different manner of knowledge of nature from that practiced in modern science. Modern science recognizes its objects programmatically as *objects*, that is, as something that is accessible only through experience of the other, and not through self-experience. It is again Jonas who has pointed out that in our self-experience, we have the source of a different, excellent experience of nature. Jonas expresses this as follows: "Ourselves living, material things, we have in our self-experience spyholes, as it were, into the innermost nature of substance" (1973: 142).

It ought to be clear that with the conception of knowledge as acknowledgment, we are given a way of looking at nature that implies its ethical relevance from the outset. This does not mean, of course, that we already have such a method of knowledge at our disposal, or that it is already in any way adequately designed. It is in the following section that we will ask how such a form of knowledge might be developed.

Formative Development: Goethe

It is Goethe who, from a perceptive critique of the method of modern science, at the same time developed alternatives that can be described as an *ethically relevant* approach to nature. Goethe's alternative approaches to knowledge of nature were partly accepted into the mainstream of science—for example, as morphology—and partly marginalized—as his color theory. In formulating the question of whether there exists an epistemological approach to nature that already implies an ethical attitude toward nature, it is worthwhile today to study Goethe's beginnings once again, and to test them for the possibility of continuing from them. The key concept that unites knowledge of nature and ethics for Goethe is the concept of *education (Bildung)*. In 1806/7 Goethe planned to publish his various works on morphology, which finally happened only ten years later in the series of *Morphological Notebooks*. In the course of publication short introductions were added, which took as their theme the very type of knowledge Goethe had in mind. In the text *The Project Introduced* (Goethe 1987) we read:

> If we become aware of natural objects, and particularly living objects, in such a way that we wish to gain an insight into the interplay of their being and function, then we believe that we can best arrive at such knowledge by means of separating their parts; as indeed it has suited us to follow this very path a great distance. We can with but a few words remind friends of knowledge of that which chemistry and anatomy have contributed to our insight and overview of nature (p. 391)

In this passage Goethe openly acknowledges the achievements of modern science, moreover in the field of the life sciences. Nowadays, in addition to the above-named disciplines of chemistry and anatomy, we could mention physiology, biochemistry, and even genetic technology. Goethe now combines this acknowledgment, however, with serious reservations:

> But these efforts at separation, carried on again and again, also produce many a disadvantage. The living thing is dissected into its elements, to be sure, but one cannot from these put it together again and quicken it. This is true even of many anorganic, to say nothing of organic bodies (ibid.)

In his acknowledgment of the scientific method, Goethe had at first named only the one part, namely analysis. In fact, of course, synthesis is also part of the method. Science is discovery by means of reconstruction. However, Goethe opines, in the area of living things one cannot reconstitute the object out of its elements—namely because after the analysis, it is dead. This is a peculiar and irritating argument. In the first moment one is quite willing to grant Goethe this argument. To be sure, one cannot reassemble a living being, once anatomically dissected, and bring it to life. But is synthesis as an element of scientific discovery really meant so concretely? Must it always be as in Lavoisier's famous experiments, for example, which proved that water is a compound (Lavoisier 1930)? He dismantled water by reducing it over red-hot iron and then finally synthesized it by converting the hydrogen into *breathable air*. Does synthesis always have to be so concretely possible? Or is the theoretical insight into synthesis sufficient? Goethe obviously believes that this insight is impossible. It is impossible to understand life from the interplay of its dead elements. That could only be possible if living creatures were some kind of machine. When a machine is stopped, its functioning can be perceived very easily from the natural interplay of its parts. Goethe does not consider this possible with living beings and therefore demands a different type of knowledge. Since he even faces up to the fact that this type of knowledge, under certain circumstances, must also be demanded for inorganic processes, he obviously has in his mind's eye a type of object not limited to the phenomenon of life I suspect that the particular quality of this type of object lies in its temporal form.

In the analytical scientific method it is assumed not only that everything appertaining to an object is contained in its elements, but also that each object, at some point in time, completely exists. The basic idea of this assumption can be grasped particularly clearly in mechanics. There this idea states that the entire sequence of movements of a body is potentially contained in its momentary state. In this respect, philosophically, force or momentum has been understood as a representation of movement (Leibniz 1982 [1695]: sec. 5). Goethe, in comparison, holds that there are objects, in particular living beings, that are what they are not at all in the moment, but rather only in performance. He considers something similar to be possible in inorganic objects too, in principle, and nowadays we can perhaps follow him along this line all the way to the background of chaos theory. But let us remain with the living beings.

Living beings, being what they are, are never in a single moment completely in existence; they are, as it were, smudged over time. They have, to draw an analogy to a different temporal entity, the character of a melody. A melody too, after all, cannot be represented in the moment by that which it is at the time, namely by the single note. To understand the melody, one must follow it through time. The biologist and natural philosopher Christoph Rehmann has recently worked out this thought with the aid of the Aristotelian concept *of praxis* (Rehmann-Sutter 1996). He criticizes the contemporary life sciences because they always characterize processes of life only in regard to their achievements, and this means that organisms are described by means of their functions. Crucial in determining what life is, claims Rehmann, is not what it produces—what Aristotle called *poiesis*—but rather the performance of life itself, or in Aristotelian terms, *praxis*. Rehmann's characterization of life, gained through a reconstruction of Aristotle, leads to the same result as I formulated in my dismissal of the paradigm of physics. Life cannot be represented by something that is given at any one point of time—neither by a force nor by a result, because it is what it is essentially only in its entire span of performance.

Goethe, however, characterizes the fully lived lifespan as such by his term *Bildung*. For our formulation it is particularly significant that by means of this term he can articulate the sense of relatedness between the performance of life and its knowledge, namely through simultaneous development:

> Thus, too, the man of science has at all times been distinguished by a drive to recognize living developments [*Bildungen*] as such, to grasp their outwardly visible, tangible parts in their interconnection, to accept them as intimations of their inner workings and so to master the whole, as it were, by means of contemplation. How closely this scientific desire is connected to the artistic and imitative drives, I trust need not be elaborated upon, (ibid.: 391)

For Goethe the point is "to recognize [*erkennen*] living developments [*Bildungen*] as such." We know that he understands these developments to be a lifelong metamorphosis that has a certain pattern, and which also extends beyond the existence of the individual through the reproduction of the genus. This thought is worked out most elegantly in the *Metamorphosis of Plants*, and best just where the relationship between scientific desire and the artistic and imitative drives mentioned in our quotation comes to the fore, namely in the eponymous poem, Here Goethe shows that the knowledge of living developments is acquired through empathetically going with them; that is, that the knowing observer, in the study of life processes, simultaneously experiences himself as a living thing.[3] Here we have the relationship that we also encountered in the ideas of Herder; namely, that in this type of knowledge, knowledge of nature is at the same time self-knowledge, to the extent that we ourselves are nature. This kind of knowledge is therefore also a formative development in this sense: the formative development of the knowing subject—through the study of nature.

The type of knowledge of nature that Goethe practices and considers particularly necessary for the study of life is thus never an indifferent determination of fact, but rather a participatory consummation. The subject of knowledge changes in the cognitive process itself; it becomes fully aware of itself in the course of studying nature. Methodologically, this thought is most clearly expressed by Goethe when he names the part of his color theory that actually presents the theory not as "Theory," but rather "Didactics," Goethe's writing thus presents itself not as a textbook, but as instructions for self-fulfillment.

Knowledge of Nature as Self-Knowledge

We have proceeded from the assumption that in our time the relationship between humanity and nature has become a moral question—and that is as much as to say: our decision determines how we understand ourselves as human beings and what kind of society we want to live in The manifold ethics that nowadays deal with our relationship to nature, be it to nature external to us or to the human body, that is, to the nature that we ourselves are, bear witness to the deep insecurity that our unbalanced relationship to nature has aroused in our own conception of ourselves. These multifarious ethics—bioethics, animal ethics, medical ethics, ecological ethics, etc.—are left dangling, as long as there occurs no change in our conception of nature itself. As long as we maintain that what nature *is* is determined exclusively by modern science, then we will be dealing with a nature that is not ethically relevant for us; nature, whether it be external nature or the nature that we are, remains superficial to us, an object, a machine, a system, a physical thing. It therefore requires the development of an *ethically relevant knowledge of nature* (Altner et al. 2000). Such knowledge, however, must be developed from an attitude that is already implicitly an ethical attitude. Such an attitude is described by Hans Jonas as responsibility, by Herder as an acknowledgment of nature, by Goethe as *Bildung,* and by Rehmann as simultaneous development with the life of nature. We have seen that from this attitude, a knowledge of nature can be developed that essentially makes use of the fact that we ourselves are nature, and experience in and of ourselves what nature is. This type of knowledge is connected to a direct, living experience of nature, but has never been developed systematically. It stands out in its main characteristics, however, in the above-mentioned authors and, at least in the work of Goethe and Adolf Portmann, has been developed to some extent and tested in concrete studies of nature. We are faced today with the task of carrying on this work, for without a different knowledge of nature, an ethics of nature will grab at thin air and remain ineffective.

Conclusion

We have demonstrated a great deal, and yet at the same time it is also too little. In a world whose prejudices are scientistic through and through, and in which, if need be, external direction of the life sciences is conceivable, yet alternative sciences are not, it is already a great deal not only to have demonstrated the scientific-theoretical possibility, but indeed in two examples to also have shown its reality. On the other hand, the references to such scientists as Goethe or Adolf Portmann are references to scientific approaches that were marginalized even in their authors' lifetimes. Against the possibility of doing things differently there stands the actual power of the established business of science, and in particular its integration into strategics of economic exploitation. Even were many bioscientists convinced of the possibility of an ethically relevant knowledge of nature, the question still arises: With what material forces could their ideas be combined, in order to have a real chance of being accepted? This would probably only be possible if this other knowledge of nature could make *curative promises* comparable to the established life sciences. It would have to be shown that alternative approaches to the knowledge of life promise possibilities of a *salutary* treatment of birth, performance of life, and death, comparable to the manipulative life sciences and biotechnologies, and indeed with lower costs in terms of the destruction of life—all the way from animal experiments to iatrogenic diseases. This does not seem so hopeless. An analogy may make this clear: had we invested from the beginning in alternative forms of energy, especially solar energy, the same means as were applied to fostering nuclear energy, then today the alternative forms of energy would long since have become competitive. In order to create an analogous situation for the life sciences and thereby for a future medicine, to be sure, a *mobilization from within*—that is, a forceful movement among the scientists themselves—would be required. And whence should this come, if not as a reaction to the growing pressure to legitimacy that the bioscientists find themselves subject to? Does this not lead back to regulation from without, namely public opinion, legislation and politically influenced allocation of means?

Notes

1. The theme of respect has been made the central point of an ethics of nature in Paul W. Taylor (1986).
2. The anthropocentric principle plays a role in cosmology (Breuer 1984).
3. I provide a corresponding interpretation of the poem in *Für eine ökologische Natunästhetik* (1999).

References

Altner, Günter, Gernot Böhme, and Heinrich Ott, eds. (2000) *Natur erkennen und anerkennen. Über ethikrelevante Wissenszugänge zur Natur*. Kusterdingen: Die Graue Edition.

Böhme, Gernot (1991) "Das Tier in der Moral," *Merkur* 45: 344–7.

—— (1999) *Für eine ökologische Naturästhetik*. 3rd edition. Frankfurt am Main: Suhrkamp.

—— (2001 a) "Die Vernunft und der Schrecken. Welche Bedeutung hat das genetische Wissen: Naturphilosophische Konsequenzen," in *Was wissen wir, wenn wir das menschliche Genom kennen?* Ludger Honnefelder and Peter Propping (eds.). Cologne: DuMont. 189–95.

—— (200lb) *Ethics in Context. The Art of Dealing with Serious Questions*. Cambridge, U.K.: Polity Press.

—— (2002) *Philosophieren mit Kant. Zur Rekonstruktion der kantischen Erkennt-nis- and Wissenschaftstheorie*. 2nd edition. Frankfurt am Main: Suhrkamp.

Breuer, Reinhard (1984) *Das anthropische Prinzip. Der Mensch im Fadenkreuz der Naturgesetze*. Frankfurt am Main: Ullstein.

Goethe. Johann Wolfgang (1987) *Sämtliche Werke*. 1st edition. Volume 24. Frankfurt am Main: Deutscher Klassiker Verlag.

Heidegger, Martin (1957) *Sein und Zeit*. 8th edition. Tübingen: Niemeyer.

—— (1983 [1929/30]) *Grundbegriffe der Metaphysik. Welt, Endlichkeit, Einsamkeit*. Lectures 1929/30 of *Gesamtausgabe*. Frankfurt am Main: Klostermann.

Herder. Johann Gottlieb (1799) *Reason and Experience: A Metacritique of the Critique of Pure Reason*. Leipzig: J.F. Hartknoch.

Jonas, Hans (1973) *Organismus und Freiheit*. Göttingen: Vandenhoeck & Ruprecht.

—— (1988) *Das Prinzip Verantwortung*. 8th edition. Frankfurt am Main: Insel.

Kant, Immanuel (1790) *Kritik der Urteilskraft*. Berlin: Lagarde und Freiderich.

Lavoisier, Antoine Laurent (1930) *Das Wasser*. Leipzig: Wilhelm Engelmann.

Leibniz, Gottfried Wilhelm (1982 [1695]) *Specimen Dynamicum*. Part I. Hans Günther Dosch, Glenn W. Most, and Enno Rudolph (eds.). Hamburg: Meiner.

Meyer-Abich, Klaus Michael (1997) *Praktische Naturphilosophie. Erinnerung an einen vergessenen Traum*. Munich: C.H. Beck.

Rehmann-Sutter, Christoph (1996) *Leben beschreiben. Über Handlungszusammenhänge in der Biologie*. Würzburg: Königshausen und Neumann.

Seel, Martin (1991) *Eine Ästhetik der Natur*. Frankfurt am Main: Suhrkamp.

Spaemann, Robert, and Reinhard Löw (1991) *Die Frage Wozu? Geschichte und Wiederentdeckung des teleologischen Denkens*. 3rd edition. Munich: Piper.

Taylor, Paul W. (1986) *Respect for Nature*. Princeton, N.J.: Princeton University Press.

Wolf, Ursula (1990) *Das Tier in der Moral*. Frankfurt am Main: Klostermann.

2

Traditional Knowledge in Modern Society

Wolfgang van den Daele

There may be numerous definitions for "traditional" knowledge; all of them will imply some demarcation from the concept of scientific knowledge. For the sake of this paper I define traditional knowledge as

- "embedded knowledge": that is, knowledge that conveys not just information, but that also has social and cultural meaning and gives the holder of such knowledge a sense of belonging and certitude;
- "embodied knowledge": that is, knowledge that cannot be represented adequately in explicated rules or textbooks, but is ingrained in people through socialization and the incorporation of skills and habits.

This definition implies that scientific (natural) knowledge can be characterized as both disembedded and disembodied. I propose to specify this characterization using the classic indicators of objectivity; universal validity, and accessibility through communication (publication). I am aware that philosophers and sociologists have taken great pains in recent decades to dismantle—or, if you prefer a more fashionable term —to deconstruct these items as myths (see Knorr Cetina 1995). However, philosophers and sociologists have gone too far in this respect, and they are beginning to lose sight of the reality of science. Analytically, at least, one can distinguish a more holistic concept of *knowledge as culture* from a more reductionist concept of *knowledge as information.* The former concept puts the emphasis on context and considers knowledge as integrated with and bound to overall systems of belief and social practices. The latter concept of knowledge puts the emphasis on differentiation and considers knowledge as culturally "special," implying that it emerges from specific types of learning and that knowledge claims will be subjected to knowledge-specific tests of validity. It seems plausible to expect that knowledge as culture will be local (community-related) and personal, and that knowledge as information will be global (valid across community boundaries) and impersonal I will associate the former with tradition knowledge and the latter with modern (natural) science.

With these distinctions in mind I proceed in four steps. First, I will specify some elements of the differentiation of modern scientific knowledge by recalling how it emerged from the background of existing culture in the history of western societies. Second. I will point out the forces of instrumentalism that have been released in modern societies and explain why the notion of objective natural knowledge resonates so effectively with the existing social institutions. Third, I will examine in what sense modern knowledge, while disembedded and disembodied in principle, is still contingent upon local contexts and personal skills. A final section gives arguments why efforts to bridge the gap between modern and traditional modes of natural knowledge are bound to fail and why the tension between science and ethics cannot be overcome.

The Differentiation of Objective Knowledge—
The Separation of the Natural from the Sacred

By the end of the seventeenth century, the inquiry of nature had largely become committed to the "edge of objectivity" (Gillispie 1960) in European societies. Claims of natural knowledge had to be based on observation and experiment, and not just on argument; causal explanation was considered the only proper principle of understanding. The notion of objective knowledge set the assertion of facts categorically apart from religious belief, moral rule, and political order. As a modern philosopher-physicist put it: Natural knowledge was reduced to "a description of the possibility and of the outcome of experiments—a law of our ability to call up phenomena" (von Weizsäcker 1960: 173). Such knowledge is the core of science. Strictly speaking, it cannot qualify as truly value-free or politically neutral, since it embraces an operational concept of truth, which deeply reflects an interest in prediction and control. However, given that bias, objective knowledge is universally valid. The predictions of modern science come true whatever the social valuation or the political order may be.

There was a cultural price to be paid, so to speak, for this universalism. Objective knowledge deprives the natural cosmos of moral and religious qualities. It severs the ties between the sacred and the natural. In scientific terms, the order of nature is devoid of deeper meaning, it no longer conveys existential certitude, it is of no help when we strive to understand why we live, how we should live, what are the right things to do, and what is our place in this world. Using a famous phrase of the German sociologist Max Weber, one may speak of the "disenchantment of the world" (Weber 1984 [1919]) For Francis Bacon, speaking at the beginning of the seventeenth century, nature was already nothing but a "storehouse of matters" from which humans might take whatever they want. The suggestion that "the inquisition of nature is in any part interdicted or forbidden" was dismissed by him as mere superstition (1860–64: 225–29). Bacon was ahead of his time. Historically, the establishment of objective natural knowledge was a matter of extensive cultural and political controversy throughout the seventeenth century and so was the separation of the natural from the sacred. I will give a few examples.

The unity or, at least, compatibility between natural and religious truth was a major issue in the debates over Copernican astronomy that waged from the late sixteenth century onwards. However, considerable differentiation already underlay these debates. First, the participants were, to a certain degree, professional scholars, intellectual specialists operating in social arenas of intellectual work at a distance from the religious life of ordinary people. Second, within the intellectual sphere, theologians and astronomers had different roles in terms of their training and competence. In the case of Galileo, the astronomers of the church accepted Galileo's claim that moons circulate around the planet Jupiter, but they handed the subsequent problem of how such a finding could be integrated into the religious dogma back to the theologians. Such division of labor is very different from the identity of astronomy and priesthood, which, for example, is ascribed to the early Babylonians.

Nevertheless, compatibility of the new astronomy to prevailing religious beliefs was imperative Andreas Osiander, in his foreword to Copernicus' *De revolutionibus* tried to shield the clash of astronomy with religious truths off by assigning scientific explanations the status of mere "hypotheses" that account for the phenomena ("provide a calculus consistent with the observation") but "cannot in any way attain to the true causes" (Copernicus 1992 [1543]). Osiander's proposal is very modern; it anticipates the instrumentalist conception of scientific theory. The proposal was not easily accepted by his contemporaries: neither Bacon nor Copernicus, nor Galileo nor Descartes were content to just "save the phenomena." Nevertheless, the direction was set: The truth claims and the spheres of validity of natural and religious knowledge had to be disentangled.

Yet, the scientific movement in the seventeenth century was unde-cided and ambiguous about this solution. The so-called chemical philosophers envisaged a fusion of the advancement of learning and piety. Chemistry was considered a key to both the natural and the supernatural. On the one hand, experiments counted as methods to achieve spiritual enlightenment, and, on the other, religious devotion was seen as the prerequisite and ultimate basis of true natural knowl-edge (Debus 1970, 1975). However, this holistic notion of knowledge that would be both natural and religious was considered heresy by the established churches and got the chemical philosophers into trouble. The mechanical philosophers who finally determined the definition of the new- knowledge replaced it by a less ambitious—and less dangerous—vision of the science—religion relationship. For instance, in 1665 Robert Boyle praised the "excellency of theology" and acknowl-edged that in cases of conflict humans should forego new knowledge rather than deviate from the duties of Christianity (see Boas Hall 1965: 142). Actually, however, Boyle left no room for such conflict. He claimed that the pursuit of natural knowledge by revealing the works of God would inevitably make the researcher a devoted and admiring believer. This was a remarkable shift. While, for the chemical philos-ophers, spiritual involvement was a prerequisite, a reason for natural knowledge: for the mechanical philosophers, it was a consequence. This was an elegant way to emancipate natural knowledge from religious belief. An intrinsic connection between science and religion was still accepted, but it would no longer constitute or restrict the contents of knowledge. This emancipation was further backed by the notion that God had created nature perfectly and endowed it with immutable laws, and that he normally would not interfere with what he had created (Daston 1997). If this is so, the laws of nature can be recognized without reference to God—even by non-believers.

The Power of Science and the Winds of Change in Western Societies

At the end of the seventeenth century the concept of positive, objec-tive knowledge was adopted by most natural philosophers; sacred knowledge and the profane were clearly differentiated. However, these philosophers were still marginal in terms of the power structure of the society at that time. Early science made its peace with the established cultural and political forces by emphasizing its indifference. In his draft charter for the Royal Society of London in 1663, Robert Hooke promised that the new society would confine itself to improving natural knowledge and mechanical arts through experiment, "not meddling with Divinity, Metaphysics, Moralls, Pollitics, Grammar, Rhetorick or Logick" (see Ornstein 1963 [l938]: 118). Apparently, such restraint was necessary to win the approval of King James II, who had just restored the monarchy after the turmoil of the Cromwellian revolution. Moreover, it also truly reflects the peculiar political neutrality of modern science. Scientific knowledge defies any commitment to specific moral, religious, or political meaning. As Eric Weil insisted, such meaninglessness has itself meaning: scientific knowledge is (instrumental) power (Weil 1965). However, this is the kind of power that can serve any power and, in that sense, it is neutral.[1]

The power of objective knowledge was merely a notion, an intellectual project, or a vision at that time. It was power *in abstracto*. The real social functions of the new natural philosophy were close to nil. It played no significant role in education or technology. When modern science finally became a social force, it was because cognitive progress had been achieved within science, but, more importantly, also because modern societies had undergone major transformations that had a profound influence on the role of and the attitude towards innovation.

Science can only contribute to technology, if the ways that things are done can be changed in a society. This is not as self-evident as it seems for a modern observer. Societies may literally encode their technical knowledge and competence in the established practices and defend these as essential ingredients of the existing social order. The medieval guilds are a good example of just such a case. While artists were allowed to change their techniques and select from among whatever they could find (that is why the "liberal arts" were so called) ordinary craftsmen had to apply techniques that were canonized through professional tradition and legal rule. Such policy was to protect the community from the disruptive forces of competition and uncontrolled social change. To that end, technical knowledge was to be kept within the community and not traded away or transmitted freely. The policy was not particularly successful in suppressing innovation—even in the Middle Ages. But it meant that innovation was considered to be—and accordingly sanctioned as—a subversive activity, that is, socially and politically deviant.

This assessment was completely reversed with the advent of the bourgeois society. Emerging capitalism "disenchained the productive forces," to borrow the phrase from Karl Marx. It turned technological innovation from an underground current into a legitimate goal and deliberate strategy. Social change through innovation was no longer considered as a threat to the existing social order, it was part of that order, thus demarcating the transition from a static to a dynamic image of society. Historically, processes have been complicated and slow. We have seen and still see many efforts by governments and social movements to contain the dynamics of technological innovation through political intervention. However, with the differentiation of the capitalist economy and the concurrent evolution of a legal culture of individual (private) rights, a bias towards innovation has been built into the society. It is not the change, but the control of change that is in need of special justification. Thus, there is legitimate space for instrumental activism to unfold in the economy and in the professions in general. "Adaptive upgrading" (Parsons 1966) or the growth of "material culture" (Ogburn 1966 [1922]) is a built-in trend and legitimate goal in modern societies.

Under these conditions, it seems pointless to consider an existing state of the art of technical knowledge as a communal order that could be reserved for the community and preserved in order to protect it against disruption. Efforts of governments or trade unions to ward off innovations that transform established professions and devaluate the personal knowledge employees have acquired through investment in education have typically failed. A late example was the strike British trade unions organized in the printing industry to obstruct the introduction of electronic printing in the 1970s. The only thing they achieved was that old printing houses were shut down and electronic businesses opened next door.

In modern societies there are no protective belts of social and cultural values to preserve the status quo of technology. In this sense, technical knowledge has also become disembedded. To avoid misunderstanding, technology is clearly embedded in the sense that social forces determine its direction and use, and moral values restrict applications. However, no constraints operate in modern societies that attempt to petrify the status quo of technical practices. This is a clear difference to instrumental practices inscribed in the immutable order and rituals of traditional societies. Modern technical knowledge is information that may be replaced easily or moved from one social context to another, and that is susceptible—you may also say: vulnerable—to scientific change. Historically, this notion of technical knowledge emerged before the scientific revolution. It is exemplified by numerous works like

De re Metallica by Agricola (1566) that registered existing (traditional) techniques and presented them in textbooks as recipes, that is, as information. Thus, while science may count as the paradigm of modern knowledge, the earlier and more basic process was the modernization of technical knowledge.

In general terms: The winds of change in Western societies are based on processes of differentiation that precede the growth of science. The natural philosophers of the seventeenth century set the inquiry of nature culturally apart from religious, moral, and political commitments and paved the way for the intellectual pursuit of the growth of knowledge for its own sake. Capitalism provided and still provides the major opportunity structure, so to speak. It imbued society with an insatiable thirst for innovation and made the economy the field that would readily absorb new scientific knowledge and transform it into social practice—the demands of the nation state (mainly for military technology) and the chronic need to improve professional (medical) performance being other scientific inroads into the society.

Personal and Impersonal Knowledge and the Indispensable Practitioner

If knowledge is reduced to information it becomes impersonal. The possessor of the knowledge does not count. Knowledge as information is "disembodied" unlike knowledge as skill. The latter are personal and "embodied," for instance, in the hands of a craftsperson who knows how to repair shoes or in the intuitions of a teacher who knows how to address a rebellious class. Information is valid and valuable irrespective of the person who provides it. It can be stored, printed, reproduced, and multiplied through communication. Published scientific results are the prime example of such knowledge. The results belong to nobody and virtually to everybody. Whoever can read the publication has access to the knowledge it contains and to the technological potential it implies. Hence, an article from a California laboratory can trigger research in India, and a blueprint for a computer written in Japan can be read (and executed) in Germany.

This picture image of impersonal, disembodied, and global knowledge needs some modification. Even in science, digesting the information communicated in a publication does not necessarily mean that one *knows* what is published—if knowing implies that one knows how to do the science, that is, reproduce the experiments reported, use the results, etc. To that end, theories, data, explanations, algorithms, rules written on paper, or disseminated through the Internet are not enough. Personal knowledge (Polanyi 1964) is necessary; this must be acquired through socialization and a local context or culture that is conducive to the kind of knowledge that is published (instruments, orientations, standards). Strictly speaking, knowledge is universal only to the extent that the personal competencies and the local circumstances that produce (or reproduce) that knowledge are universal.

In the field of technical knowledge the amount of know-how that cannot be explicated in information may even be higher. Contexts under which techniques operate (function) are closer to the real world complexities than the idealized laboratory conditions under which scientific experiment takes place. Patent applications must disclose information that would enable those familiar with the existing state of the art to copy the new technology. In reality, such information is seldom sufficient. Licenses to use the patents are usually combined with contracts that transfer know-how, and this may imply additional information or equipment, or providing persons who can run the technology. This example proves a more general point: Know-how cannot be fully converted to information, technology is more than applied science (in contrast to Bunge 1966), and information abstracts from context. To be useful, however, the information must function in context. It is the knowledge (the skills) of the practitioner that recontextualizes information to make it operational, in modern engineering it may be the scientist who devises a new machine, but it is the mechanic in the workshop who actually runs it.

Recourse to the knowledge of the practitioner may be imperative even in modern technology. This, however, does not reintegrate the technology into social practices shared by the "people." The gap between the scientist and the practitioner is a gap between two professional cultures. Mechanics are experts and as far as their knowledge is concerned, they are as distant from ordinary lay people as they are from the trained scientist.[2]

The growth of modern science has in many fields replaced the implicit and embodied knowledge of the practitioner by the explicit and disembodied knowledge of the scientist. The dynamics of invention had until well into the nineteenth century been a domain of the practitioners; most pioneers of the industrial revolution experimented (and thus applied some of the methods of research but they found little help from scientific theory). Since then science has taken the inventive lead. Scientific knowledge is covering increasingly complex subjects.

How far that can go is difficult to predict. In dealing with complexity, the limits of the knowing scientist may be narrower than the limits of the knowing practitioner, for instance in handling human life and behavior, organizations, or technical systems. But note that computer recording and simulation is already effectively "reading" the skills of craftsperson and translating them into machine programs for auto-mated factories. This opens up new ways to disembody the knowledge of the practitioner and reduce it to impersonal information that can be universally used—even without explicating and understanding the underlying rules of the practice.

Science and Ethics: Essential Tension and Irreversible Rift

Modern natural knowledge, both scientific and technical, is informa-tion and as such culturally disembedded, which means: differentiated from encompassing moral values, social concerns, and political goals. Obviously, knowledge is being developed with a view to social, eco-nomic of political purpose. This is evident with technology, but even the path for theory building in science can be shaped by such purposes if the respective capacities have been reached in a discipline (see Böhme et al. 1983). However, the options implied in the knowledge developed for a purpose are more general than the purpose. Medical knowledge can be used to kill. Research in the making of an atomic bomb provides insight to build a nuclear power plant. In that sense, instruments—and instrumental knowledge—are neutral. They may be devised for a certain context, but they can be transferred to other con-texts—if the values and rules enforced in the society allow such transfer. This "if" is essential. It is by no means the case that technical options arising from scientific knowledge will automatically be realized. Not everything that can be done, will be done. However, restrictions come as external constraints: legal duties, professional codes, economic interest, etc. The constraints operate on the application of knowledge, not on the validity or the truth of knowledge. Thus, even where moral concerns clearly prevail over instrumental options, as with certain high risk interventions or manipulation of humans, they still prove the disunity of knowledge and ethics in the modern (Western) culture.

It is somewhat hard to accept the implication of this finding since it confirms the conventional professional ideology of science claiming the moral neutrality or even innocence of the knowledge and putting eventual blame on the decisions made in the society of how to apply the knowledge. So far, however, no operational proposal has ever been made to transcend the conventional solution. Even if scientists could be persuaded (and sanctioned) to observe a new Hippocratic oath that they provide knowledge only for the sake of humanity and the common good (or sustainable development, for that matter), such an oath would be an external control on the application of knowledge, not an intrinsic feature of their knowledge. Whatever they know would still be information that could be converted to other uses, which then must be ruled by I he society. Thus the essential tension between knowledge and ethics is reasserted and unresolved.

Is there a chance to go beyond or back behind the notion of knowledge as information in modern societies? After all, we do have more holistic notions in our culture. We have knowledge of values, needs, and desires. We know love and death. We know (at least sometimes) what really matters in human life. Such knowledge is meaningful beyond informational contents. Such meaningful ness is extended to things, places, and events that might otherwise be described in objective terms, and makes them, as Michel Leiris put it, traces of the "sacred in everyday life" (1938). These are traces of the sacred in a secular society. We also have such traces in our relationship with nature. It would be a caricature of modern culture to suppose that its notion of nature is defined by natural science and in fact coincides with the Baconian "storehouse of matter." You may consider your own relationship with landscape, animal life, weather, or water and detect that we relate to nature also in admiration, fear, veneration and, in a sense, communication! The modern predicament is not that the deeper meaning of nature and the perception of the sacred in things have become inaccessible in out culture—we still know them. The predicament is that such knowledge is dissociated from and external to the instrumental knowledge of nature accumulated through science and technology.

And there is little hope of ever bridging this gap and integrating (or reintegrating the different ways of understanding nature. The notion of objective positive science has often been contested. But the dream of an alternative natural science has never come true. As Gillispie observes, "Its ruins lie strewn like good intentions all along the ground traversed by science" (1960: 199). Ecology seems to be the latest victim. Some decades ago Gregory Bateson hailed ecological thinking because it would restore a "sense of unity of the biosphere and humanity which would bind and reassure us all with an affirmation of beauty" (1980: 19). The expectation, however, that such restoration could be achieved within science was frustrated.

Ecology did not turn out to be a "subversive science" (Shepard and McKinley 1969). It is not the alternative to the Baconian vision of nature as a resource and of natural knowledge as instrumental power. Scientific ecology conceives of nature as an evolving, interconnected self-regulating system. And this is indeed a different image than the 'clockwork' universe of classical physics. But this image does not imply the transition to a non-instrumental, perhaps contemplative or moral view of nature. Rather it is the transition to a more sophisticated technological view. Francis Bacon insisted that nature could only be commanded by obeying its rules. If ever, it will be through the understanding of ecosystems that humans will really be able to command nature by obeying it. Scientific ecology does not transcend the Baconian vision of the domination of nature, it perfects it (see also Cramer and van den Daele 1985).

But let us assume for a moment that a new natural knowledge can be devised that restores the sense of the unity of the biosphere and humanity and includes the affirmation of beauty. Would that knowledge be able to replace the objectivism and positivism in the established order of knowledge in science and technology? Very unlikely! Rather it would be another variant of knowledge in our culture. It would add to the pluralism and heterogeneity of types of understanding that exist: scientific, religious, moral, aesthetic, personal, contemplative, biographical, etc. It might even find a niche in the university curriculum, as did feminist studies. But it would not replace mainstream science and technology and its focus on objective natural knowledge. Such knowledge is irresistible in modern societies not because it is the only feasible notion of true knowledge (which may be impossible to prove), but because it is congruent with the instrumentalism engrained in these societies. This pattern is not going to be reversed. As long as we have a value system that backs individualism and individual rights, as long as we have a differentiated capitalist economy, and as long as we presume that political intervention must solve remaining problems, resources will be defined as chronically scarce in the society. If it is not the scarcity of food, or health, or security, it will be the scarcity of material products in general or of the power to control. The way nature is dealt with in the economy and the professions may be inadequate in many respects. Modern societies will have to learn to control their control of nature if they are to meet human needs in a sustainable manner, preserve the planet, and do justice to future generations. But this is still a quest for control and adds to the demand for knowledge that promises control.

Instrumental knowledge is itself a resource that is chronically scarce in modern societies.

Therefore, and this is my final remark, whenever the agents of modern societies confront traditional cultures that may have more comprehensive embedded notions of natural knowledge, they are likely to screen such knowledge for useful information and appropriate the information into the modern framework, disregarding the deeper meaning the knowledge has in the culture of origin. Whether traditional communities can preserve that meaning once they come into contact with modern societies remains to be seen. They may also adopt the modern instrumentalist view and redefine the traditional knowledge as a resource for medical, agricultural, or environmental techniques, that is, reduce it to knowledge as information. The latter seems to be implied, if they pursue the protection of traditional knowledge under a regime of intellectual property in order to reserve the right to commercialize such knowledge and forestall 'bio-piracy' by modern industries (see, e.g., Nijar 1996).

Notes

1. If modern science acts as a servant of the political order, it is indirectly providing the material benefits of new technology. Otherwise, it is politically compatible by way of indifference, not through the symbolic reflection and intellectual foundation of the existing order. Of course, "not meddling" may also be considered a contribution to political legitimation. A further contribution could be that concepts of science provide suitable metaphors for political philosophy. Shapin and Schaffer (1985) claim that this is the case in the relation between Boyle and Hobbes.

   As a servant of the political order, science may be attacked if power is reversed. Thus, the French Revolution dismantled the scientific institutions of the Ancient Regime and killed some prominent representatives, only to rebuild institutions and re-establish the scientists later (Hahn 1971).

2. The same holds for the shepherds Brian Wynne places vis-à-vis nuclear scientists in his account of the controversy over when sheep could graze safely on English hillsides after the Tscherobyl fallout (1996). Not surprisingly, the shepherds knew better in many respects. Nevertheless, this does not prove that lay wisdom has a place in scientific matters; it shows that the expertise of scientists is limited to their model systems and experiments but that the expertise with respect to the behavior of sheep rests with shepherds.

References

Agricola, Georg (1556) *De Re Metallica*. Basel: Froben.

Bacon, Francis (1860–64) *The Works of Francis Bacon.* Volume IV. James Spedding. Robert Leslie Ellis, and Douglas Denon Heath (eds.). Boston: Drown and Haggard. (Reprinted [1963] Stuttgart: F.F. Verlag.)

Bateson, Gregory (1980) *Mind and Nature: A Necessary Unity.* New York: Bantam.

Boas Hall. Marie (1965) *Robert Boyle on Natural Philosophy: An Essay with Selections from his Writings.* Bloomington: Indiana University Press.

Böhme. Gernot, Wolfgang van den Daele. Rainer Hohlfeld, Wolfgang Krohn, and Wolf Schäfer (1983) *Finalization in Science: The Social Orientation of Scientific Progress.* Volume 77 of Boston Studies in the Philosophy of Science. Wolf Schäfer (ed.). Dordrecht: Reidel.

Bunge, Mario (1966) "Technology as applied science," in *Technology and Culture* 7: 329–47.

Copernicus, Nicholas (1992 [1543]) *On the Revolutions.* Edward Rosen (ed. and trans.). The Johns Hopkins University Press.

Cramer, Jacqueline, and Wolfgang van den Daele (1985) "Is ecology an 'alternative' natural science?" *Synthese* 65: 347–75.

Daston, Lorraine (1997) "The nature of nature in early modern Europe," *Configurations* 6(2): 149–172 (Berlin: Max Planck Institut for the History of Science. Preprint 59.)

Debus, Allen G. (1970) *Science and Education in the Seventeenth Century: The Webster-Ward debate.* London: MacDonald.

—— (1975) "The Chemical Debates from the Seventeenth Century: The reaction to Robert Fludd and Jean Baptiste van Helmont," in *Reason, Experiment and Mysticism in the Scientific Revolution.* Maria Luisa Righini Bonelli and William R. Shea (eds.). New York: Science History Publications. 19–47.

Gillispie, Charles Coulston (1960) *The Edge of Objectivity: An Essay in the History of Scientific Ideas.* Princeton: Princeton University Press.

Hahn, Roger (1971) *The Anatomy of a Scientific Institution: The Paris Academy of Sciences, 1666–1803.* Berkeley: University of California Press.

Knorr Cetina, Karin (1995) "Laboratory Studies: The Cultural Approach to the Study of Science," in *Handbook of Science and Technology Studies.* Sheila Jasanoff, Gerald E. Marke, James C. Peterson, and Trevor Pinch (eds.). Thousand Oaks, Cal.: Sage Publications. 140–66.

Leiris, Michel (1979 [1938]) "Le Sacré dans la Vie Quotidienne," in *Le collège de Sociologie (1937–1939).* Denis Hollier (ed.). Paris: Gallimard. 60 ff.

Nijar, Gurdial Singh (1996) *In Defense of Local Community Knowledge and Biodiversity. A Conceptual Framework and the Essential Elements of a Rights Regime.* Third World Network Paper 1. Penang, Malaysia.

Ogburn, William Fielding (1923) *Social Change: With respect to culture and original nature.* London: Appleton.

Ornstein, Martha (1963 [1938]) *The Role of Scientific Societies in the Seventeenth Century.* Reprinted from the 3rd edition. London: Archon.

Parsons, Talcott (1966) *Societies: Evolutionary and comparative perspectives.* Engelwood Cliffs. N.J.: Prentice Hall.

Polanyi. Michael (1964) *Personal Knowledge: Towards a post-critical philosophy.* 2nd edition. New York: Harper and Row.

Shapin, Steven, and Simon Schaffer (1985) *Leviathan and the Air Pump: Hobbes, Boyle, and the experimental life.* Princeton: Princeton University Press.

Shepard, Paul, and Daniel McKinley, eds. (1969) *The Subversive Science: Essays Towards an Ecology of Man.* Boston: Houghton Mifflin.

Weber, Max (1984 [1919]) "Wissenschaft als Beruf," in *Gesammelte Aufsätze zur Wissenschaftslehre.* 5th edition. Tubingen Mohr. 582–613.

Weil, Eric (1965) "Science in modern culture, or the meaning of meaninglessness," *Daedalus* 94: 171–89.

von Weizsäcker, Carl Friedrich (1960) *Zum Weltbild der Physik.* 8th edition. Stuttgart: Hirzel.

Wynne, Brian (1996) "May the Sheep Safely Graze? A reflexive view of the expert–lay knowledge divide," in *Risk. Environment and Modernity: Towards a New Ecology.* Scott Lash, Bronislaw Szerszynski, and Brian Wynne (eds.) London: Sage Publications. 44–83.

3

In Search of Vehicles
for Knowledge Governance:
On the Need for Institutions
that Creatively Destroy
Social Capital

Steve Fuller

It does not make sense to speak of the governance of knowledge without specifying knowledge-bearing institutions. Nevertheless, both the philosophy and sociology of science have been traditionally lacking in this regard, which in turn has generated a knowledge-society problem comparable to the classic mind-body problem. As a corrective, I discuss some knowledge-bearing institutions, the ideal case of which is the university.

The university is distinguished by its engagement in what I call the creative destruction of social capital. That is, in their research function, universities create advantage, in its teaching function, they destroy it. This dual function has been historically tied to the university's institutional autonomy. However, as the university has incorporated more of society into its activities—and thereby truly universalized the knowledge it produces—it has opened itself to factors that threaten to dismember its institutional integrity. Although affirmative action is sometimes portrayed by its detractors in this light, a much stronger factor is the incursion of capitalist modes of production, which I call capitalism of the third order. This tendency, while accelerating with the decline of the welfare state, has had many historical well-wishers, who together reveal liberalism's instinctive scepticism toward knowledge-bearing institutions combined with an openness to information technology. Moreover, as the state has shifted its role from provider of knowledge as public good to regulator of intellectual property, a curious rewriting of the politics of knowledge governance has occurred. Thus, much of the critical thrust of my paper focuses on the influential claim by Edmund Kitch that knowledge tends to escape its bearers, unless the state arrests its flight through legislation. Because the exact opposite is truer to history, the significance of the university as a knowledge-bearing institution tends to be grossly underestimated.

The Governance of Knowledge: Some Semantic Preliminaries

When first dunking about the phrase, "the governance of knowledge," we should compare it with "the governance of breathing." The latter refers to a process that is normally treated as part of the "autonomic nervous system" of individuals. In other words, it is a discipline that the body self-imposes without conscious thought. Classical empiricist epistemologies would seem to treat the governance of knowledge in much the same way, at least if we take empiricist talk of reality impressing itself on the mind at face value. In this context, conscious thought typically appears as fixed ideas that block or distort the mind's natural receptiveness to reality's impressions, much as blocked blood vessels may impede the flow of oxygen. However, empiricist epistemologies have come to be rejected on two grounds. One is that conscious thought seems to enter the governance of knowledge in a much more positive sense; the other is that the governance of knowledge appears to occur at a collective, not an individual, level. These two points imply that if knowledge is regarded as something that can be governed, then it must be formally constituted at a collective level, which is to say, clearly distinguished from other activities that individuals do for themselves, perhaps automatically. In short, our breathing may be governed individually and unconsciously, but our knowledge is governed collectively and consciously. This, in turn, implies the alienability of knowledge from those who would claim to possess it, a concept that figures prominently in the pages that follow.

But we should also attend to the use of "governance." This word may be usefully contrasted with "management"—or, for that matter, "government"—in terms of whether the principal and the agent of control are the same. In governance they are, in management and government they are not. Historically, managers were a class of experts distinct from the owners (typically shareholders) of the corporations for which they worked. Thus, the recent coinage "knowledge management" implies, among other things, that neither the producers nor the con-sumers of knowledge are necessarily well suited to control its flow. Some third party formally outside the system is better placed to do so. I regard Nico Stehr's original locution for our project, "policing knowledge," as evoking an extreme version of this perspective, in its suggestion that an ill-managed knowledge system might transgress ordinary societal norms. In contrast, governance implies the autonomy of the system under consideration, and hence an identity of the producers and the governors of knowledge. It follows that if knowledge is not seen to be governed properly, then the knowledge producers themselves need to take the appropriate course of action. This may mean that they open up their ranks and/or revise their norms. But either strategy will be presented as self-legislated, not externally imposed. My defence of the institution of the university shall presuppose "the governance of knowledge" in this sense.

As these semantic preliminaries already suggest, sensible talk about the governance of knowledge requires an identification of the relevant institutional vehicles of knowledge production. This turns out to be harder than one might first expect because most normative talk relat-ing to knowledge, including science, is conducted without reference to specific institutional contexts. For example, judged by the breadth of his examples, Robert Merton's (1973) famous account of the normative structure of science was meant to apply to institutions as disparate at the seventeenth-century Royal Society of London and twentieth-century U.S. "Big Science." This traditional indifference to institutions by sociological—and not just philosophical and economic—analyses of knowledge helps explain the shock value of work in the recent social studies of science, which has made a point of highlighting knowledge's physically embodied and socially embedded character. In this respect, traditional talk about the governance of knowledge suffers from many of the same problems as a normative psychology or a theory of rationality that is severed from discussions of brain science and human physiology, more generally. In short, questions of the governance of knowledge dwell in a knowledge-society split at least as old and problematic as the divide between mind and body (Fuller 1993a; Fuller 1993b: chap. 9).

Following the pattern of the mind-body problem, there is a "cheap" solution to the knowledge-society problem. The scholastics called it multiple instantiation but nowadays the solution travels under the more modern name of functionalism. It claims that that the same idea may be conveyed by many different brain states or, for that matter, computer stales, just as the same proposition may be communicated in many different sentences, in many different languages. By analogy, then, the same knowledge—say, of atomic physics—may be produced in many different institutional settings: a university, a corporate science park, or a government laboratory. In each setting, the researchers' roles and the resulting social dynamics will also be different. Yet, somehow these disparate social practices are supposed to contribute to a common body of knowledge. While such a view seems implausible when stated so baldly, it is nevertheless taken for granted in most science policy contexts. Indeed, it is enshrined in the definition of "science indicators," which are concerned exclusively with characteristics of papers published in scientific journals, regardless of the institutional origins of their authors.

What makes the above solution to the knowledge-society problem "cheap" is that it regards the governance of knowledge purely from the *consumer's* perspective, thereby reducing it to a knowledge-management problem. This point may not be so apparent when a philosopher of mind says that various brain and/or computer states express functionally equivalent thoughts. However, when a philosopher of language observes that various sentences express functionally equivalent meanings, the question is soon asked whether some of these sentences might not express the same meaning more clearly, which is to say, more efficiently from the interpreter's standpoint. Similarly, a knowledge manager armed with science indicators may argue that, say, universities and science parks produce functionally equivalent knowledge of, say, biotechnology. But which arrangement produces this knowledge more efficiently and, if so, what policy lessons may be drawn? The knowledge manager does not consider that these knowledge producers, precisely by virtue of their different institutional settings, may embed their work in incompatible intellectual contexts and perhaps not even see themselves as engaged in the same enterprise. Of course, this point cuts both ways: It not only undermines the homogeneity, of what normally passes for a body of knowledge, but also inhibits knowledge producers from checking the advance of knowledge management by organizing themselves into "knowledge worker" unions (Fuller 2002a: chap. I).

The Missing Term: Knowledge-Bearing Institutions

In short, without explicit attention to the institutional vehicles of knowledge governance, debates over whether knowledge producers should govern themselves or be subject to external control are bound to be radically underdetermined in terms of policy implications. I shall look briefly at two divergent German proposals along these lines, the first by Gernot Böhme, the second by Wolfgang van den Daele. It is worth noting that both had been prominent in a previous life in the so-called finalization movement, which in the 1970s wedded a largely autopoietic conception of scientific change (inspired by Thomas Kuhn) to a state-led strategic research initiative for "mature" sciences. (An English translation of the main papers of this movement is Schaefer 1984.) Nowadays Böhme and van den Daele would seem to emphasize, respectively, the former and latter stages of the original finalization vision.

Böhme has argued that the scientific community should look to its own history to find an ethically sounder, more ecologically conscious version of itself, perhaps grounded in the holistic worldview of nineteenth-century *Naturphilosophie*. In this respect, contemporary public mistrust of science is symptomatic of science's own self-alienation. While there is much to commend in Böhme's perspective, it leaves open the question of institutionalization. Presumably, the plan here runs deeper than simply exhorting mature scientists to dip into the works of Goethe for inspiration.

One can imagine at least two possible policy implications that would radically change the face of knowledge production. The first would be to require that science students study the histories of their fields, so that they learn of alternative frameworks for casting their inquiries before they have been indoctrinated into the dominant framework. The second would be to involve a broader range of scientific practitioners into the governance of their own fields, since it is well-known that the more distant scientists see themselves from cutting-edge specialist research, the more open they are to interdisciplinary and extradisciplinary (a.k.a. transdisciplinary) concerns. However, both policy implications would encounter stiff resistance from the scientific establishment for what followers of Mary Douglas in the sociology of scientific knowledge would recognize as "group-grid violations" (e.g., Bloor 1983). On the one hand, scientists would have to open their doors to approaches in opposition to which their professional identities were originally forged (i.e., a group violation); on the other, they would have to subvert science's meritocracy, which values researchers in terms of their closeness to specialist research frontiers (i.e., a grid violation).

But perhaps all this shows is that the problems facing science's relationship to the larger society will not be solved through new self-governance policies. In that case, we have van den Daele's proposal that science's applications be subject to external control, especially so as not to interfere with the promotion of alternative, so-called indigenous knowledges. This proposal presupposes that some rather clear distinctions can be drawn between science and other forms of knowledge, as well as between research and its applications. These presuppositions should recall the knowledge-society problem discussed above, and the interpretive ambiguity of its exact policy implications: Are we to be concerned with the promotion of certain forms of knowledge, regardless of who the knowledge-bearers are, or the promotion of certain groups of people, who now bear certain forms of knowledge but need not do so in the future?

In short, van den Daele's proposal does not squarely face the Enlightenment challenge that people will improve their lot in life by transferring their allegiances from traditional to more scientific forms of knowledge. In other words, just because knowledge is historically tied to a particular group, or "stakeholders," it does not follow that such knowledge remains indefinitely relevant to promoting their lives. Now, of course, certain forms of indigenous knowledge may be worth cultivating to complement, temper, or revise scientific knowledge. But whether that task is best executed by indigenous knowledge-bearers is an open question. A good benchmark for these issues is the practice of linguists interested in preserving minority languages in the face of globalized anglophonia: Insuring a steady supply of native speakers is not the only possible strategy. Non-natives could be also encouraged to learn the minority languages.

As we have seen, both Böhme's and van den Daele's proposals for governing knowledge underestimate the normative import of institutions. Both are primarily oriented to articulating a normative vision—Böhme's of an updated *Naturphilosophie* and van den Daele's of a state-enforced cultural pluralism—that leave open the exact institutions that might realize them. However, if one is interested in having these outcomes recur on a regular basis, and not simply through ad hoc interventions, then the selection of appropriate institutional vehicles is of paramount significance. Normative solutions to problems of epistemic legitimacy may appear authoritarian if exclusive attention is paid to articulating the ends of knowledge production, thereby implying that virtually any means is justified in the process of achieving those ends.

Moreover, even when institutional frameworks are given serious thought, sometimes the ones chosen fail to produce a whole greater than the sum of the original parts. In that sense, the institutions may reflect the sociological presuppositions of their designers without enabling their participants to accomplish more than they would, were they left to their own devices. Examples may be found in van den Daele's own highly publicized "technology assessment" panels, which bring together major interest group representatives—as stakeholders are less delicately called—to design policy guidelines for such controversial issues as genetically modified foods (van den Daele, Pühler, and Sukopp 1997). Stalemate often results, which in turn has been used to justify a government-based ethics advisory board lo sort out irresolvable value differences in concrete situations. The problem here, I submit, is not that the different interest groups have radically different world views, but rather that the technology assessment panels are designed to reinforce, not sublimate, those differences. Thus, cultural pluralism becomes a self-fulfilling prophecy.

For an institutional contrast, consider the "consensus conference," which I have defended as vehicles for public involvement in setting science and technology policy (Fuller 2002a: chap. 4). This set-up would restrict the role of stakeholders to witnesses in a trial, the jurors for which would be members of the public who are not formally affiliated with any of the stakeholders. Of course, the citizen jurors may already enter the proceedings with strong views, but their livelihoods would not be directly affected by the prospect of changing their minds over the course of the consensus conference. The sociological presupposition here is that, on any given policy issue, most people are not so clearly and strongly aligned with a particular worldview that they are impervious to argument. The much decried "ignorance" of the public on matters of science and technology is a weakness *only* if it results in closed-mindedness. But the appeal to a broad range of interest group representatives as witnesses in consensus conferences is specifically designed to reveal the highly contested nature of the knowledge that the public allegedly lacks, which in turn forces conference participants to be both more vigilant and open-minded, as well as encourages them that they can "make a difference."

Of course, none of this guarantees that particular citizen-juror solutions will completely satisfy the stakeholders. But here the institutional legitimacy of the consensus conference would prevail over objections to its particular decisions. This is easier said than done, however, since European governments (not least the United Kingdom) increasingly include consensus conferences as only one of many public consultation devices on science and technology policy matters. Unfortunately, this pluralization of institutions can be tantamount to no institution-alization, if the state is permitted maximum discretion in its use of these devices. What the public says depends very much on how it is addressed, and multiple addressings virtually ensure contradictory responses that can be then played off against each other, as political expedience requires.

Thus, to count as an improvement over stakeholder panels, consensus conference outcomes must be binding on science and technology legislation, while at the same time allowing for the revision of that legislation in the future, according to its consequences. Moreover, because these are general normative desiderata for the design of institutions, they raise questions about the ultimate desirability of, say, maintaining strong distinctions between scientific and indigenous knowledges, if the relevant knowledge-bearers manage to shift their epistemic orientations after their participation in consensus conferences. Indeed, much of what is diagnosed as antagonism toward science may really be alienation from the science policy-making process, which over time can be reified, with the help of intellectuals and other suppliers of myth-making materials, into a "positive" alternative epistemic identity, a so-called imagined community (Anderson 1983). Consensus conferences are well-positioned to reverse that tendency. However, historically speaking, one institution has been unique in its capacities for governing knowledge by simultaneously reproducing and delegitimating it. This is the university, to which we now turn.

The University as the Ideal Knowledge-Bearing Institution

In the time-honored equation "knowledge is power," power involves *both* the expansion and contraction of possibilities for action. Knowledge is supposed to expand the knower's possibilities for action by contracting the possible actions of others. These others may range from fellow' knowers to non-knowing natural and artificial entities. This broad understanding of the equation encompasses the interests of all who have embraced it, including Plato, Bacon, Comte, and Foucault. But differences arise over the normative spin given to the equation: Should the stress be placed on opening or closing possibilities for action? If the former, then the range of knowers is likely to be restricted; if the latter, then the range is likely to be extended. After all, my knowledge provides an advantage over you only if you do not already possess it. In this respect, knowledge is what economists call a positional good (Hirsch 1977), a concept that will loom large in the pages that follow.

In this context, it helps to explain our rather contradictory attitudes toward the production and distribution of knowledge. We do research to expand our own capacity to act, but we teach in order to free our students from the actions that have been and could be taken by others.

By virtue of their dual role as producers and distributors of knowledge, universities are engaged in an endless cycle of creating and destroying "social capital," that is, the comparative advantage that a group or network enjoys by virtue of its collective capacity to act on a form of knowledge (Stehr 1994). Thus, as researchers, academics create social capital because intellectual innovation necessarily begins life as an elite product available only to those on the cutting edge. However, as teachers, academics destroy social capital by making the innovation publicly available, thereby diminishing whatever advantage was originally afforded to those on the cutting edge. Recalling Joseph Schumpeter's (1950 [1942]) definition of the entrepreneur as the "creative destroyer" of capitalist markets, the university may be regarded as meta-entrepreneurial institution that functions as the crucible for larger societal change. This process mimics the welfare state's dual economic function of subsidizing capitalist production and redistributing its surplus. Not surprisingly, then, universities magnified in size and significance during the heyday of the welfare state, and have been now thrown into financial and wider institutional uncertainty with the welfare state's devolution (Krause 1996).

Moreover throughout its history, the university has been institutionally predisposed to engage in the creative destruction of social capital. In the Middle Ages, they were chartered as permanent self-governing bodies in a world of limited sovereign reach. Keeping the peace was often the most that a realistic sovereign could hope to achieve. Thus, in exchange for loyalty to the local ruler, universities were legally permitted to set their own curricula, raise their own capital, and even help manage the region's everyday affairs. This was the context in which universities were chartered as among the first corporations (i.e., *universitates,* in Medieval law). This orientation marked a significant shift from the much more populous residential colleges of the Islamic world, the *madrasas,* which depended on the benefaction of intrusively pious patrons, or the more venerable, but also more routinized, training centers for civil servants in imperial China (Collins 1998). To be sure, like these institutions of higher learning, the Medieval universities were broadly dedicated to the reproduction of the social order. However, because the universities were founded in times and places that were profoundly *disordered*, academics were immediately thrown into situations where their words and deeds effectively brokered alternative futures.

Given these origins, it is not surprising that academics have found it relatively easy to seed social unrest, which invariably they have interpreted as bringing order to an otherwise disordered situation. Perhaps the signature case of universities' imposing order is the Humboldt-inspired research-and- teaching university of the modern era, which is fruitfully conceptualized as a social technology for incorporating large segments of the population into the production and distribution of knowledge (Fuller 2002b). For example, exemplary works by eccentric geniuses were transformed into employment schemes for ordinary trainee academics. Kuhn would later call this routinization the "disciplinary matrix" sense of "paradigm," which has become the backbone of modern graduate education (a.k.a. normal science). Thus, modern academia transformed Newton's *Principia Mathematical* from an imperfectly realized masterwork to a blueprint for a collectively realizable project. More generally, this attempt to cast the university as social technology for truly universal knowledge has accelerated the institution's tendency to drift from what I have called a monastic to a priestly mode (Fuller 2000a: chap. 5; Fuller 2002a: chap. 4): the former stressing the virtues of institutional autonomy, the latter those of societal transformation.

Perhaps the clearest epistemic marker of this drift is the benchmark for original research. In the monastic mode, the inquirer's empirical resources are typically confined to the university's grounds, which means a reliance on the campus library or oneself (or sometimes students) as primary databases. Under the circumstances, historical and philosophical studies provide the *via regia* to knowledge of the particular and the universal, respectively. But as the university has extended its political ambitions into the priestly mode, these two disciplines were replaced, respectively, by sciences focusing on ethnographic field work and experimental laboratory work. Accordingly, universities have undertaken substantial commitments to transform and govern areas, or "sites," often far off-campus. This has not only driven a physical and psychological wedge between the university's teaching and research functions, but it has also recast the university as a participant in power structures about which many of its staff, over the years, have had serious reservations. Yet, at the same time, staff loyalty to particular universities has diminished, so that nowadays complainants are more inclined to look toward the greener pastures of other campuses than to try to reform their current institution.

However, the most obvious recent university policy that illustrates the university's priestly mission is affirmative action legislation, which quite explicitly takes forward the university's regulative ideal of creatively destroying societal advantage by giving priority to traditionally underprivileged groups in the hiring and promotion of academic staff as well as the selection and sometimes even evaluation of students (Faundez 1994). This point, which generally goes unappreciated by the policy's many critics, highlights the distinctive sense in which universities (and other chartered corporations) have participated in the more general processes of societal reproduction. For, here we have a legally self-perpetuating social institution whose process of inter-generational role replacement is not family-based. In other words, *universities are pioneers in the decoupling of social reproduction from biological reproduction.*

In the United States, the country with the most developed affirmative action policies, there are two independent justifications for the practice, one relating to private and the other to public sector institutions. In private sector institutions, the justification recalls the founding of the early universities as church-based institutions: No one is fully responsible for their lot in life, since, in the words of the old Lutheran hymn, "there, but for the grace of God, go I." This attitude, associated with a vivid sense of the inscrutability of divine justice, or "theodicy," motivates not only academic administrators but also successful alumni to campaign to increase the ranks of their home institutions with recruits from disadvantaged backgrounds. Thus, the very high tuition fees of the Ivy League are, for most students, subsidized through alumni contributions. This enables the attendance of poor but worthy students, who in turn become the next generation of alumni to proselytize the university's lifelong virtues.

In contrast the public justification for affirmative action has been much more controversial because, in effect, it involves turning the universities into the institutional vanguard of a more democratic future, a role that the university's traditionally elite status makes it ill-equipped to perform (Bell 1966). It is one thing for the university to facilitate the social advancement of those capable of succeeding on strictly academic terms, but quite another for the university to reverse generations of both earned and unearned differences in social status—especially when tuition fees are kept artificially low by taxes that, generally speaking, are paid in proportion to the likelihood that the taxpayer's family will *not* benefit from such a reversal. Thus, an indefinite policy of affirmative action immediately raises questions of the reciprocal obligations of inter-generational justice. Arguments for inter-generational justice gain their intuitive purchase from potential benefactors and recipients sharing sufficiently similar values (if not more genetically based properties) that, say, our descendants would thank us for whatever steps we take now to protect the natural environment. But why should I enable hostile minorities to enjoy the benefits of university attendance, if neither they nor I would recognize each other as worthy benefactors and recipients?

However, the public justification of affirmative action becomes even more controversial, once its purview is narrowed from the open-ended promotion of distributive justice to the more limited charge of commutative justice. Here universities would be used for only a fixed period to redress past damages done to the opportunities of specific groups to advance their interests in society. This policy seems to suppose that the ideal university would embody a vision of societal equilibrium, with just the right proportion of the relevant groups represented in the composition of the faculty, student body, curriculum, etc. Such a view is controversial for two reasons. It presupposes, first, that specific social groups are marked as being distinctly advantaged or disadvantaged, and second, that each group carries a mind-set or worldview that cannot be adequately represented by anyone from any other such group (Fuller 2000a: chap. 4). These two presuppositions constitute the arena in which the so-called canon wars have been fought on U.S. university campuses for nearly two decades. Here what detractors call political correctness codes would divide the liberal arts curriculum into major works by authors associated with specific socio-cultural groups, in proportion to their numbers in the national population—and not by universally applicable standards of critical judgment.

Interestingly, the U.S. courts have been more inclined to overrule affirmative action legislation in cases that have sought to redress specific past damages than to promote a general vision of the future. The major problem with enforcing commutative justice many years after the fact is that it is difficult to identify the descendants of the originally aggrieved parties, not least because the constitution and interrelationship of the relevant social groups have changed over time. This is typically presented as a practical problem in the administration of justice, but it points to the deeper theoretical point that knowledge tends to escape its original social setting—but not necessarily because it becomes universally available (contrary to a thesis of Edmund Kitch, to be discussed in the next section). Indeed, this gap between the fugitivity and universality of knowledge creates a space for an institution like the university, which renders the fugitive universal by increasing the ranks of legitimate knowledge-bearers. In a similar spirit, U.S. courts may look more favourably on distributivist claims for affirmative action because they help promote the ultimate civic republican virtue—freedom of expression as more of the population is both exposed to a wide range of views and provided with the material conditions (i.e., job skills) for deciding between those views *with impunity*.

This last point implies two distinctive, but often neglected, systemic features of universities. The first is that, contrary to much recent thinking on the topic, the imperatives of liberal education and vocational training need not work at cross-purposes. Instead, they may be mutually reinforcing, precisely because they are separate but equal, which makes it possible to acquire the knowledge needed to earn a living without first having to commit to a particular worldview. In other words, you can change your religion or political party affiliation, and still keep your job (or at least find comparable work elsewhere). This aspect of modern academic instruction should not be underestimated as a vehicle for instilling independent-mindedness. Nevertheless, the events of 11 September 2001 demonstrated its flipside, since the "Muslim extremists" who crashed into the World Trade Center and the Pentagon learned their flying skills in the West, without first having to commit to such liberal values as respect for the rule of law.

The second neglected systemic feature is that universities are, strictly speaking, more concerned with *redistributing* than *reproducing* the social order, since academic merit is defined independently of the usual class markers. Admittedly, there may be a strong contingent link between intellectual and economic status, but academics have traditionally regarded this more as a problem than a brute fact. Put more abstractly, universities routinely adopt standards that question how the past is to be carried into the future, as people from both advantaged and disadvantaged backgrounds come to do well by those standards. Moreover, it follows that for any given generation, those who have already achieved academic success would have themselves come from a variety of political and economic backgrounds. This fact serves to institutionalize internal division among each generation of academic elites, who then have reason to recruit potential allies from the larger society, a pattern that is familiar from the historic extension of democratic rights to the entire population from its civic republican and aristocratic origins (Moore 1963).

In assessing the argument of this section, an uncharitable reader may conclude that I have resorted to a purely "functionalist" analysis of the university that fails to address the actual effects of its so-called creative destruction of social capital. More concretely, has increasing the ranks of academics with women and ethnic minorities led to marked changes in the forms of knowledge that are valued, or simply an admission that a wider range of people are capable of institutionalizing the dominant forms of knowledge? For example, Barbara McClintock objected to Evelyn Fox Keller's (1983) feminist account of her work in maize genetics because McClintock saw herself as having succeeded by the standards of scientific inquiry upheld by less-than-obliging "old boys," not standards that were liberalized to allow a specifically women's form of knowledge to flourish. Something similar may be said of the Harvard Afro-American Studies Department spearheaded by Henry Louis Gates (1988), who has been much keener to stress racially inspired neglect of superior Black literature than the need to revise literary standards to accommodate Black contributions.

It would seem, then, that the lesson so far taught by affirmative action is that white males have failed to live up to their own universalist aspirations, not that universalism as conceived by white males is a false ideal. Perhaps unsurprisingly, women and ethnic minorities are now *over*-represented in senior administrative posts in U.S. universities in exactly the sort of bloodless transfer of power that the Enlightenment *philosophes* originally desired—but failed to realize in the French Revolution of 1789. For, contrary to the more heated rhetoric of the campus canon wars, affirmative action has *not* led to a radical transformation in the kinds of knowledge pursued in the academy. To be sure, since the 1960s, social movements devoted to the promotion of specific ethnic, class, and gender identities have gained academic footholds. But they quickly adopted the university's preoccupations with the production, reproduction, and evaluation of specialist forms of knowledge that typically owed their theories and methods to more established disciplines (The tendency of feminist theory to become applications of, say, Foucault, Lacan, or Habermas, provides a vivid example; Nicholson 1994.) In that sense, the priestly mission of the university has been enormously successful in colonizing heretofore non-academic fields. Readers can decide for themselves whether this has amounted to co-opting or empowerment of the people that these fields purportedly represent.

Nevertheless, the university is currently undergoing some significant changes in the forms of knowledge it produces. But affirmative action has had little to do with this situation. Rather, the worldwide devolution of the welfare state has opened universities to market forces, which have pressed for the "accountability" of academic knowledge with more efficacy than socialist-inspired appeals to "relevance" in the 1960s and 1970s. Of course, accountability is now understood in terms of clients and consumers, whose claims matter in proportion to their spending (and sometimes voting) power. In that case, efficiency measures—value for money—acquire an overriding normative significance in the evaluation of research and education policies. With an irony that only an Orwell can truly appreciate, many of the effects of this "new production of knowledge" travel under such old progressive rubrics as "transdisciplinary" and "problem-oriented" (Gibbons et al. 1994). However, whereas the original progressives merely succeeded in expanding the scope of traditional academic knowledge, today's knowledge capitalists have gone quite far in assimilating research to intellectual property and education to employment credentials. I shall now argue that this development can be seen as capitalism rendered "self-conscious," or as I say below, "capitalism of the third order."

The Knowledge Society as Capitalism of the Third Order

There are two general ways of thinking about the nature of capitalism. The more familiar one is a first-order account about how producers are engaged in perpetual—and largely self-defeating (according to Marxists)—competition to make the most out of the least, and thereby generate the greatest return on investment (a.k.a. profits). Whatever its other merits, this account takes for granted that the relative standing of competing producers is self-evident, so that no additional work is required to identify the market leaders. But in fact, such work *is* needed. This second-order account of how producers publicly demonstrate their productivity is the context in which "capitalism" was coined by Max Weber's great German rival, Werner Sombart, in 1902 (Grundmann and Stehr 2001). What contemporaries, notably Thorstein Veblen, derided as the "conspicuous consumption" of successful capitalists, Sombart treated as the principal means by which capitalists displayed their social standing in a world where social structure was no longer reproduced as a system of fixed heritable differences. Thus, capitalists had to spend more in order to appear more successful.

However, it would be misleading to think of these expenditures as allowing capitalists to luxuriate in their success. On the contrary, it spurred them to be more productive in the ordinary, first-order sense, since their competitors were quickly acquiring comparable, if not better, consumer goods. Indeed, before long, the competition was so intense that it became necessary to spend on acquiring the connoisseurship needed to purchase goods that will be seen—by those who know how to see—as ahead of the competition's purchases. By the time we reach this third-order capitalism, we are at the frontier of the knowledge society.

A certain vision of economic history is implied in the above account of capitalism. In pre-capitalist times, consumption was done at the expense of production, which explained (for example) the fleeting success of Spain and Portugal as imperial powers. They failed to reinvest the wealth they gained from overseas; they simply squandered it. In contrast, capitalist consumption is second-order production supported on the back of increased first-order production. From a sociological standpoint, the most striking feature of this "before-and-after" story is its suggestion that capitalism is innovative in altering the sense of responsibility one has for maintaining a common social order. In pre-capitalist times, this responsibility was, so to speak, equally distributed across its members, regardless of status. Lords and serfs equally bore the burden of producing the distinction that enabled lords to dominate serfs. Expressions like "mutual recognition," "respect," and "honor" capture this symmetrical sense of responsibility. However, in capitalist times, it would seem that, like insurance in today's devolved welfare states, individuals bear this burden in proportion to their desire to be protected from status erosion. Thus, those who would be recognised as superior need to devote increasing effort to a demonstration of their superiority.

This last point becomes especially poignant in advanced capitalist societies, where at least in principle the vast majority of people can lead materially adequate lives while spending less time and effort on first-order productive pursuits. However, as the political economist Fred Hirsch (1977) famously observed, this situation simply leads people to intensify their efforts at second-order pursuits. As a result, for example, individuals spend more on education and firms on advertising, even though the advantage they gain in terms of first-order production is marginal or temporary. Yet, this expenditure is necessary to be seen as "running with the pack." The logic of what Hirsch called the production of positional goods predicts that, over time, one's relative status will decline, unless it is actively maintained, which usually involves trying to exceed it, thereby raising the absolute standard that everyone needs to meet. Thus, an expanded production of positional goods, combined with increased efficiency in the production of material goods, results in the systemically irrational outcomes that we have come to expect (and perhaps even rationalise) as our knowledge society. Specifically, the resources spent on acquiring credentials and marketing goods come to *exceed* what is spent on the actual work that these activities are meant to enhance, facilitate, and communicate.

Of course, such a classic case of means-ends reversal is *not* systemically irrational, if it marks a more-or-less conscious shift in values. Thus, it may not take much to be persuaded that we really do produce in order to have something to sell, and we take up particular jobs in order to have a platform for showing off our credentials. The *struggle for recognition* therefore overtakes the *struggle for survival:* The ultimate triumph of the German over the English tradition in political thought (Fukuyama 1992: chaps. 13–19; cf. Fuller 2001). But this point acquires more of a sting in the case of what economists call public goods, especially knowledge. In the case of such goods, producers are (supposedly) not only unable to recover fully the costs of production, but they would also incur further costs, were they to restrict consumption of their goods. However, I would urge that so-called public goods should be analyzed as simply the class of positional goods that most effectively hide their production costs.

But before uncovering the hidden production costs of public goods, a few words are in order about an influential, but ultimately wrong-footed, attempt to capture knowledge production's distinctive struggle for recognition—namely, in terms of the practice of "gift-giving," as anthropologists since Marcel Mauss have characterized non-commodity forms of exchange in so-called primitive societies (Gregory 1994). This attempt, associated with Mertonian sociologists of science, especially Hagstrom (1965), the first monograph to treat science as a community. The animating idea of gift-giving is that tribal members are recognized for the amount of themselves or their wealth they give away indiscriminately. While the recipients of this largesse are not indebted to the gift-giver in the sense of having to match the size of specific gifts, if they do not themselves provide gifts of a similar nature over time, their own status diminishes. The normative analog from science is sufficiently clear: Scientists publish their work, not to increase their income, but to impress their peers, who may nevertheless be better placed to develop that work. Pierre Bourdieu (1975) began to chip away at the model's plausibility when he observed that highly competitive gift-giving could have the same destructive consequences as any other unregulated market. His ironized concept of symbolic capital thus comes very close to Hirsch's positional good.

However, a deeper objection to the gift-giving model of epistemic recognition is that it mistakes effect for cause. Specifically, the model suggests that scientists are recognized in proportion to the amount they contribute to their discipline's collective knowledge base, when in reality it is the recognition they receive from their disciplinary peers that determines the size of their contribution. In other words, a scientist is "great" not because of her magnanimity, but because of her indebtedness to colleagues who, by citing her work, donate to her a portion of their own credit. In return for these donations, the "great scientist" performs a variety of "insurance" functions for the donor. Some of these are simply performed by invoking the scientist's name, such as providing professional credibility for the donor's work in public forums. Other functions go to the heart of the political economy of scientific knowledge production, requiring the great scientist to act on the donor's behalf in support of a funding or employment opportunity. Indeed, if the scientist is already great, other researchers may place themselves at professional risk, if they do not solicit her support through an appropriate donation of credit (e.g., an article citation). In short, what the Mertonians regarded as spontaneous gift-giving by a few big scientific producers is better seen as an emergent result of protection money given by many small scientific producers (Fuller 1997: chap. 4. app.; Fuller 2000a: chap 5).

That sociologists should have found the gift-giving model even remotely persuasive speaks volumes about the long-standing alienation of science from considerations of political economy. Gift-giving is a rational form of exchange only insofar as the gift-givers are publicly recognized for the gifts they give. However, the vast majority of work published in scientific forums goes uncited and probably unread. They are gifts that remain permanently unopened—that is, unless a "knowledge manager" scavenges it from the *Science Citation Index* as part of a strategy to save a company the expense of having to do its own original research. Indeed, the strongest support for the Mertonian case comes from the more wishful elements of the self-understanding of scientists, as eloquently articulated in Max Weber's "Science as a Vocation," which portrays even the most obscure inquirer as a virtual participant in a quest that recedes indefinitely into the future.

The Welfare State's Role in Making Knowledge Appear "Self-Protective"

Nevertheless, that knowledge would be the paradigm case of a public good is itself no mystery. For example, it may have required much effort for Edison and Einstein to come up with their ideas, but once those ideas were published, anyone could potentially benefit from them. A logical conclusion of this line of thought, exploited by the U.S. legal theorist Edmund Kitch (1980), is that knowledge resists commodification to such an extent that the state must intervene to restrict its flow through intellectual property legislation, which ensures that knowledge producers can reap at least some of the fruits of their labors. Kitch imagines that knowledge is so naturally protective of its own interests that, in effect, a special class of laws is needed to protect knowledge producers from the knowledge they produce!

Thus, Edison is entitled to a patent because of the likely *commercial* benefit afforded by his ideas, since once I understand how Edison invented the first incandescent light bulb, I am in a good position to design similar goods more efficiently that can be then sold more cheaply, and thereby comer a market that would otherwise belong to Edison. (In the economic history literature, this is sometimes called the Japan Effect, whereby it is always better to run second in unregulated market competition.) But why do similar worries *not* arise in the case of Einstein's discovery of relativity theory? In other words, suppose economists took seriously both the costs of acquiring the training needed to put Einstein's theory to any sort of use and the fact that this training would allow the trainee to earn a reasonable living as a physics instructor, if not design a way to supersede Einstein's theory that would merit the Nobel Prize. It that case, questions would be raised, not only about whether Einstein might not also be entitled to some legal protection, but also whether knowledge is as naturally footloose as Kitch and other public goods theorists make it out to be.

Two interrelated issues need to be explored here. The first is the source of the difference in our normative intuitions concerning Edison and Einstein as knowledge producers: Why should the former but not the latter be entitled to legal protection? But the second, more general issue is the source of Kitch's influential intuition that knowledge is inherently self-protective. My response to the first question will lay the groundwork for answering the second question. I shall argue that by overlooking the background political economy of knowledge production, Kitch's thesis about the self-protective nature of knowledge gets matters exactly backwards. In short, specific, mostly state-based, institutions (most notably the university) have been required to ensure that knowledge possesses the sorts of properties that Kitch personifies as self-protective. It should come as no surprise that Paul Samuelson (1969), the most influential welfare state economist of the post-WWII era, coined the phrase "public good" (albeit to formalize the only non-protective function that Adam Smith prescribed for the state), or that the need for public finance schemes to support scientific research should have been first raised by a utilitarian philosopher with strong welfarist concerns. Henry Sidgwick (Lutz 1999: 110).

So let us ask: Why is Einstein *not* entitled to legal protection? Einstein's theory of relativity was innovative against a body of physical science whose development had been funded by the German state through taxation and other public finance schemes, the main beneficiaries of which were institutions of higher education. These institutions were, in turn, open to anyone of sufficient merit, who would then be in a position to contribute to this body of knowledge. Einstein happened to take advantage of this opportunity. But even if Einstein had not existed, it would have been only a matter of time before someone else would have come along to push back the frontiers of knowledge in a comparable manner—so it is assumed. But as long as it remains unclear from what part of the population the next Einstein is likely to be drawn, the public finance of higher education is justified (imagine a compulsory lottery). In that case. Einstein does not deserve the economic protection afforded by a patent because he exploited an opportunity that had been subsidized by his fellow citizens.

Now, why would the state have undertaken such a public finance scheme in the first place? Here we must resort to some political metaphysics. The state must presuppose that some knowledge is vital to the national interest, yet there is no natural incentive for any particular citizen to engage in its pursuit. Therefore, the state must provide the sort of universalized incentive scheme exemplified by free public education. Germany acquired this mindset, courtesy of Baron Helmut von Moltke, the mastermind of its victory in the Franco-Prussian War of 1870–71. Von Moltke argued that a healthy nation was always ready for "total war," that is, not merely strategic engagement with a definite goal in sight (the classical aim of warfare), but rather the ongoing removal of any threat to national security. This was the idea of a permanent state of emergency, which would come to be the signature stance toward research and education policy in Cold War America, a period of unprecedented university expansion (Noble 1991).

In a sense, then, Einstein received advance payment for the theory of relativity by having been allowed to obtain the training necessary for making his revolutionary breakthrough. To be sure, many other people underwent similar training and failed to arrive at anything of comparable significance. But that just underscores the risk that the state, on behalf of its citizens, undertakes when it raises taxes for mass public education. There is no guarantee that the benefits will outweigh the costs. In contrast, some situations that call for new knowledge are sufficiently obvious that citizens, regardless of prior training, will find it in their self-interest to try to meet them. In that case, an innovator is vulnerable to similarly oriented individuals who are in a position to make marginal improvements that end up displacing the innovator from the market. Edison's discoveries occurred in this environment, which justifies his entitlement to a patent.

Now, in either Einstein's or Edison's case, is knowledge self-protective? Clearly not in Einstein's case. On the contrary, the state had to seed opportunities for his kind of knowledge to be produced. Edison's case is a bit more ambiguous, but even here the answer is no. After all, the only people capable of capitalizing on Edison's innovation were those who were already thinking along similar lines. There is no reason to think that mass publication of the details of Edison's incandescent light bulb would have enabled most Americans to design such a product for home use, let alone mass consumption.

In order to address the more general question of the source of the idea that knowledge is somehow self-protective, I begin by returning to the eighteenth-century European Enlightenment to pose the problem in its most basic form: Should knowledge production be granted any special legal protection? What are the grounds, if any, for the regulation of intellectual property transactions—or, in less economically presumptuous terms, the regulation of intellectual life? Here *laissez-faire* and *dirigiste* responses can be distinguished

The laissez-faire response is that once people enjoy sufficient wealth not to have to live hand-to-mouth, they ought to use their leisure to improve themselves and the polity. The implied analogy, perhaps made most explicit in the opening of Aristotle's *Metaphysics*, is the imperative to physical fitness among the well-fed as a sign of both one's superior status and preparedness to defend that superiority in warfare. Moreover, one would not be capable of advancing the frontiers of knowledge, were one not in a position to expend resources on lines of inquiry that might end up bearing no fruit. Thus, the fiscal benefit typically granted to the production of intellectual innovation in the eighteenth century was a *prize*, not a salary, grant, or for that matter, royalty. In other words, the reward consisted of a largely ceremonial event to mark the formal recognition of the innovation. Potential rivals for the prize were presumed to have independent means of material support, by virtue of cither literal or adopted fathers: i.e., inheritance or patronage. (This is not the place to explore the Darwinian-cum-Freudian implications of this situation.) In either case, they harbored no expectations of living off their innovations, as today's royalty regimes potentially allow. The contest to solve some problem left over by Newton was regarded in the spirit of a game, in which even losers never lose so much that they cannot return to compete in the next battle of wits.

The dirigiste response is associated with the reasoning behind the patent law provision in the U.S. Constitution. The U.S, founding fathers, whose perspective on human nature owed more to Hobbes than Aristotle, did not believe that a free citizenry would be necessarily inclined toward the pursuit of knowledge. After all, a happy existence may be obtained through relatively effortless and unproductive means, like charging high rents to tenants on one's property. At the same time, the founding fathers also believed in the overall benefits of new knowledge to the progress of the commonwealth. This led to a characteristically eighteenth-century strategy of converting private vices into public virtue by providing explicit financial incentives for people to engage in knowledge production, namely, the temporary monopoly on inventions afforded by a patent. Moreover, since the main economic impact of a successful invention is that it destabilizes, or creatively destroys, markets, as more people seek patents, everyone else will soon have reason to engage in the same activity in order to restore their place in the market. Thus, a lethargic economy dominated by rent-seekers is quickly transformed into a dynamic commercial environment

Both the laissez-faire and dirigiste approaches to the regulation of intellectual life continue to have cultural resonance today. The idea that society is best served by individuals exercising their right to be wrong, a theme that unites civic republican democracy and Popperian philosophy of science, presupposes that inquirers are materially insulated from the consequences of their bold conjectures, just as the laissez-faire approach would have it (Fuller 2000a: chap 1; Fuller 2002a: chap. 4). More controversially, the dirigiste sensibility lurks in the "orientalism" that has led political economists from Adam Smith onward to demonize the decadence of the East in favor of the industriousness of the West, with Western aristocrats consigned to the oriental side of the divide.

A feature strikingly common to the laissez-faire and dirigiste Enlightenment approaches to intellectual property regulation is the *absence* of any assumption that knowledge is self-protective. To be sure, both approaches presuppose that new knowledge is potentially available to any rational being inclined to pursue it. However, the inclination to inquiry is not itself universal. Certain economic conditions first need to be in place before the epistemic appetite is whetted. In the dirigiste case, it consists of a financial incentive to counteract the natural tendency to gain the most pleasure from the least effort: in the laissez-faire case, it is simply a generalized cultural expectation of people who are relatively secure in their material existence.

So, if that is the view from the Enlightenment, where does the idea of knowledge as self-protective come from? As so often happens with our ideas about knowledge, the answer lies in a *syncretistic* understanding of history. That is, factors of rather different origins are treated as contributing to a common contemporary effect.

I have already indicated the determining role of what Alvin Gouldner dubbed the "welfare-warfare state" in establishing the modern political economy for the production of knowledge as a public good. Each citizen, simply by virtue of performing the fiscal duties of a citizen, contributes to the capital needed to produce public goods and, of course, becomes a potential beneficiary of that investment. However, in fact, most citizens reap modest epistemic returns from their investment, namely, the assortment of skills that enable them to cam a living. The identities of the few who benefit as Einstein did are rather unpredictable, since they would not necessarily have been in direct contact with the researchers whose work theirs builds upon or, for that matter, overturns. Rather, these innovators encounter their precursors secondhand, through textbooks and their often undistinguished classroom interpreters.

Those still in the grip of Kuhn's mythic history of science easily forget how this very basic element of knowledge consolidation and transmission—a textbook usable by the entire range of a discipline's practitioners—first emerged in the context of nation-building efforts in the late nineteenth century (Olesko 1993). In earlier times, an aspiring intellectual innovator would not have appeared credible, had he not made personal contact with a recognized master of the innovator's discipline. By such cultish means, disciplinary practitioners jealously protected their knowledge so that it could not be easily appropriated by others. And while these ancient prejudices linger in academic hiring practices, the provision of free public education has sufficiently loosened their constraint on actual intellectual innovation to leave the impression that innovators can come from anywhere, thereby contributing to the illusion that knowledge is self-protective.

Athenian Privilege as the Deep Politics of Knowledge Governance

But beyond the background assumption of the welfare state's provision of public goods lies the heavy hand of Plato, whose followers have promoted the idea that human bodies are vehicles, filters, and sometimes even impediments, to the naturally fugitive character of the thoughts communicated through them. Here we need to recall that Platonism ultimately reflects the perspective of an aristocracy that enjoyed the freedom to pursue knowledge but not the freedom to apply it as it would wish. Specifically, Plato's Academy was founded as a defensive reaction to the fall of Athens at the hands of Sparta in the Peloponnesian Wars.

The anchoring experience—or "traumatic episode"—of Platonism was that the much-wanted intellectual freedom of the Athenians, exemplified by the influence of the Sophists, resulted in a volatility that rendered the society vulnerable to enemies who did not hold such freedom in high esteem. Indeed, today we still characterize the epistemic impulse in terms of following a line of thought to wherever it may lead, which suggests that it may lead somewhere surprising and perhaps even subversive. From that standpoint, it is easy to see how one might come to believe that knowledge flow is natural and its arrest artificial, albeit a politically necessary artifice. Yet, a moment's reflection on the political economy of ancient Athens would reveal that the so-called fugitivity of knowledge is none other than the freedom enjoyed by equals in a public forum to appropriate an argument initiated to one end and turn it to another. This spirited activity, the perfection of which constituted the Sophists' trade, was after the fall of Athens virtualized into a kind of inter-generational communication among elites, so that professional thinkers now engage in what Richard Rorty, following Michael Oakeshott, pompously calls the conversation of mankind.

We should not forget that this image of knowledge's natural fugitivity would not have arisen in the first place, had the Athenian aristocrats not enjoyed the leisure to pursue knowledge as an end-in-itself abstracted from the life-contexts in which it would normally serve as a means. However, this alienation of knowledge was not restricted to self-contained wordplay in the public forum that occasionally spilled over into risky foreign adventures. The elite did not simply autonomize knowledge from everyday life. They also deployed it as a status marker that enabled them to colonize ordinary forms of labor by setting normative standards for them.

For example, when a baker baked a cake to the client's specifications, the client received credit for having conveyed the cake's design successfully to the baker and the baker received credit for carrying out the instructions as conveyed. In the former case, credit was transacted as approval by the larger peer community; in the latter, by the fee paid to the baker. In this way, a strong distinction was institutionalized between what we would now call substantive and functional rationality, or *praxis* and *techne*, as the Athenians understood the matter (Godelier 1986). Not surprisingly, Socrates could instigate public controversy by wondering aloud whether someone who could speak eloquently about baking—that is, the client ordering a cake—really knew more about baking than the baker himself. As Socrates' sophistic interlocutors made clear, the received wisdom on the matter was that, yes, the baker was little more than a prosthetic extension of his client's thoughts.

The history of guilds, professional associations, and labor unions has consisted largely of efforts to return control over the substantive rationality of work to the workers, and thereby demystify the apparently self-protective nature of knowledge. Even the organization of the medieval universities can be understood in this light, as school masters and doctors were keen to elevate their learning from the status of mere training and credentialing, the quality of which was judged in terms of their students' careers (over which they had relatively little control) rather than features of their art that only they were in a position to evaluate. As this example already suggests, the Athenian legacy has been difficult to shake off. On the one hand, universities have been continually threatened with devolving into diploma mills. On the other, many academics have happily let the quality of their craft be judged by the success of their student clientele (and, starting in the nineteenth century, their practical applications). This was also the original stance of the Sophists, who, as foreign merchants lacking full citizenship, took well to their role as high-grade bakers. That Socrates was an Athenian citizen meant that his interrogations amounted to a critique of the political economy of his own society.

More generally, the problematic nature of the Athenian legacy has triggered some of the touchstone episodes in the history of intellectual property regulation. Two in particular stand out. The first relates to the commercial aspect of publishing the second to the scaling up of the scientific enterprise. The pattern in both cases is similar: The organization of knowledge production centers on an impresario whose expertise lies in realizing an idea under appropriate material conditions. I deal with the publishing case here and the scientific case in the next section.

The modern institution of copyright originated in the late eighteenth century as a revolt against the social structure of book publishing, which up to that point had exhibited a characteristic pattern. The publisher contracted an editor who was a recognized authority in the subject matter of the proposed book. The editor's authority inhered in his knowledge of the people actually capable of writing well on the topic and his ability to persuade them to contribute to a book under *his* authorship. In other words, the writers would regard working under the editor as a status marker that would help them obtain future writing contracts. They would therefore jump at the opportunity and deliver their chapters in a timely fashion. As the Athenians would have expected, a successful editor would enhance his reputation as an authority in the field, simply by virtue of what we would now regard as his organizational skills, while the writers would receive payment for their actual work. In this context, books were quite literally manufactured "knowledge products," as opposed to, say, vehicles of self-expression.

This system had persisted for several centuries, until two trends of the early modern era conspired to bring it down. The first was that movable-type printing enabled books to be easily copied, and hence pirated; the second, that the rise in literacy across Europe swelled the ranks of people who could make a living from writing. Consequently, publishers were motivated to cut production costs, which meant not merely encouraging editors to hire people who would undertake writing assignments for the lowest fees but sometimes even eliminating editors altogether and directly commissioning single- authored texts from such writers. Modern copyright law, which protects the originality of a piece of writing, arose to distinguish writers from interchangeable manual laborers subject to Ricardo's "iron law of wages." According to the Fichtean metaphysics underlying this legislation, each writer harbored a unique genius that warranted the title "author" and hence a fair share of his publications' profits, which his name would supposedly attract (Woodmansee 1984). In practice, of course, copyright has helped to restore the epistemic integrity of writers without necessarily improving their financial situation, since a writer's reputation is not necessarily a good indicator of his works' profitability.

Updating Athenian Privilege in the Knowledge Society: Big Science

Were one to ask which institution of contemporary knowledge societies best captures the Athenian understanding of the social structure of knowledge production, the answer would have to be the "Big Science" laboratory. Already, by WWI, Pierre Duhem (1991) had identified this social form as "German science" in Wilhelm Ostwald's physical chemistry laboratory in Leipzig. Indeed. Duhem stigmatized it for the very tunnel-visioned sense of inquiry that Thomas Kuhn would come to valorize, fifty years later, as "normal science." As it turns out, Ostwald trained many of the institutional leaders of the twentieth-century chemistry, including Theodore Richards, the first American to win a Nobel Prize in the field. A student of Richards, future Harvard President James Bryant Conant, claims credit for having imported Ostwald's organizational sensibility into U.S. academia, which up to that point had been a site for the liberal education of future elites, certainly not a microcosm of the industrial division of labor existing elsewhere in society (Conant 1970: 69–72).

It is one thing for academically trained elites to extract surplus value from their employees after graduation, but quite another for academics to subject each other to this treatment, as the lineage running from Ostwald through Conant advised. Nevertheless, today's contract researchers have come to accept, for the most part, that the professor employing them should be listed first in a jointly authored publication (or grant), regardless of the actual work he or she has done (or will do) on the research reported (or proposed). This is because the professor would have provided the material conditions for the researchers' livelihoods, not least by actively campaigning for the virtues of his or her laboratory; perhaps basking in the aura of some personal research achievements from the distant past. (As in ancient Athens, the professor's rhetorical skills are not to be underestimated) Moreover, the material conditions provided by the professor extends beyond the provision of time and space. In true Athenian fashion, the professor's reputation, or social capital, both eases publication of the research and enhances a contract researcher's own ability to secure future employment. Conflicts over intellectual property rights tend to arise only when a professor prohibits researchers from using their own data in projects outside the professor's lab, and even then the conflicts are individualized as matters of research ethics (Penslar 1995).

Embodied in the social organization of the Big Science laboratory, as described above, are the two principal objections to the labor theory of value, which would have workers paid exactly in proportion to the contribution of their labor to the final product. Interestingly, these two objections pull in opposing ideological directions, from the Right and Left respectively.

Seen from the Right, contract researchers deserve *less* than what the labor theory prescribes because the professor takes most of the risks by setting up the lab, employing the researchers, in anticipation that the lab's results will attract sufficient attention from colleagues to have justified these costs. If noteworthy results fail to materialize, then the professor's reputation may suffer, but the researchers still get paid. From this perspective, an "exploited" worker is really the beneficiary of a loan from his employer that will be hopefully but by no means necessarily—repaid by consumers. This is the argument that the Austrian finance minister, Eugen von Böhm-Bawerk (1959 [1884]), used to counter Marx's theory of profit as surplus value that capitalists owed their workers. We earlier saw a version of this reasoning in the idea that Einstein was not entitled to royalties for his theory of relativity because in effect he had been already paid by the state (along with many other, less productive people) when he was granted access to as much education as his capacities permitted him.

Seen from the Left, contract researchers deserve *more* than what the labor theory prescribes because neither they nor their professors are personally responsible for their employment situation. Indeed, the observance of a sharp division between tenured and untenured academic labor is simply an efficiency measure for increasing productivity in a sealed-up knowledge economy. It is not designed to harden into a new class system. Consequently, it is imperative that contract researchers be permitted the maximum mobility that their talents allow, including the opportunity to run their own labs. Thus, professors are obliged to promote the career prospects of their workers, even if that means transferring some credit for their joint work, say, in letters of recommendation that stress a researcher's essential role in the production of noteworthy scientific results. What small amounts of personal credit the professor might lose in these acts of strategic overstatement are likely to be offset by the researcher's improved employment situation. This justification for redistributing the professor's credit on the basis of interpersonal utility comparisons (i.e., the professor's loss vis-à-vis the researcher's gain) is a version of the argument made by the Cambridge welfare economist Arthur Pigou (2000 [1920]) for a progressive income tax.

The preceding excavation of the political economy of the modern laboratory reveals that Big Science is "fiduciary" in two distinct senses, which are conflated in the practice of calling the professor the principal investigator in grant proposals. This expression immediately brings to mind the fiduciary relationship between principal and agent in business law. But exactly who is supposed to fill each role? The science policy theory literature already provides one answer, namely, the state is the principal whose entrusted agent is the scientific community (Guston 2000). However, in order for the scientific community to perform its agent role, professors must act as principals vis-à-vis the agents they employ as contract researchers to bring their grant proposals to fruition. As we have just seen, the trust needed to empower such agents results in their being accorded *both more and less* than the value of their actual labor.

Who's Afraid of Research Fraud? The View from Mars

The doubly fiduciary character of science becomes painfully apparent in what is increasingly called research fraud, which typically occurs when a professor, a researcher, or both try to improve their lab's situation by illegitimate means. The threat of corrupting the collective knowledge base merely scratches the surface of what is at stake here. More deeply, research fraud demystifies the economic ideology that has traditionally underwritten national science policies, namely, that one can never spend too much to advance the frontiers of knowledge. In the best of times, one might expect an epistemic hang proportional to the allocated bucks, but more and more will probably need to be spent to achieve less and less. Indeed, the fact that returns on basic research investment tend to diminish in the long run has been even used to operationalize Kuhn's sense of progress in a mature paradigm. At the dawn of scientometrics, Derek de Solla Price (1978) caught a glimpse of the problem here: The best predictor of a research-intensive society is electricity *consumption* per capita—not, say. Gross National Product per capita or some other production-based economic indicator.

From the standpoint of a Martian economist who does not share our (including most earthbound economists') fetishization of knowledge production, national science policy budgets look like elaborate exercises in conspicuous consumption. Moreover, our Martian friend could even find a venerable tradition of thinkers from Adam Smith to Michael Polanyi who would grant the point, especially in the case of funding academic research, though they might draw wildly different normative implications from it (Fuller 2002a: chap. 1). While Polanyi, citing Athenian precedent, would put a positive spin on science's conspicuous consumption. Adam Smith's spin would be rather more negative. His revolutionary ruminations on the wealth of nations were largely spurred by the negative example Spain set in lavishing its universities with the riches acquired from its New World conquests, while failing to develop the productive capacities of most of its people. Smith's thoughts about these matters not only echoed the antiacademic sentiments of the Scientific and Industrial Revolutions, but also resemble those of today's knowledge managers who would hold research funding to the efficiency standards of modern enterprises: that is, the *most* bang for the buck.

To the average knowledge manager, or "KM guru," successful firms increase their profit margins while financing the R&D division at just the necessary minimum. If anything, knowledge managers put a higher premium on knowledge that either captures what is already know or pre-empts what can be known, both of which are better achieved by manipulating the legal system than applying the scientific method (Fuller 2002a: chap. 2). From that standpoint, it is an open question just how problematic research fraud really is. Economically speaking, research fraud results from someone pretending to have done the work that would be normally required to reach a certain conclusion, assuming the conclusion is true. Thus, it is the researcher's pretense to effort—*not* the actual validity of her conclusion—that is mainly at issue here. Of course, the two issues are easily elided in the minds of the fraud-averse, usually by focusing on the knowledge claim's reliability, a quality in research that can be produced only at considerable cost, specifically of the sort typically circumvented in research fraud. Nevertheless, if one considers the exact knowledge claims that tend to be faked, they are often ones that the researchers believe to be true, but they simply lack the time, money, or skill to produce an appropriate demonstration.

At this point, our knowledge manager, prompted by an impatient Martian economist and propelled by the spirit of Adam Smith, might attempt to cut this normative Gordian knot by throwing the matter open to the market: Those caught in research fraud may be asserting truths or falsehoods, but why must this judgment be based on how much the researchers have *consumed* (of their own time and effort, their professors' reputation and resources, etc.) rather than how much they and other researchers can *produce* on the basis of this fraud? Indeed, the prototype of this bold question was posed—vis-à-vis Galileo's improbably idealized experiments—by Alexandre Koyre (1978 [1939]) and Paul Feyerabend (1975).

To be sure, Koyre and Feyerabend had somewhat different aims in sight Koyre wanted to reinforce the value of Platonism as a force in Western thought that has enabled the likes of Galileo to see ahead of the pack, while dragging the rest along with (faked) empirical demonstrations. For his part, Feyerabend wanted to ridicule modern philosophy of science's preoccupation with demarcation criteria and other declarations of methodological loyalty that would not be needed had charlatans (i.e., mystics, alchemists, and other purveyors of "magic") not hit on the truth so often—to be sure, by rather inscrutable and devious means. But, as in our current discussion, both Koyre and Feyerabend appealed to a sense of overall efficiency in the history of science that is not captured by what might be regarded as the science system's normative expenditure at any given moment. Koyre's solution was to cast the Platonic tradition as a semi-protected black market for intellectual risk-seekers who periodically make strategic interventions in the open market, whereas Feyerabend would simply open up scientific institutions to competitors whose very lack of the positional goods associated with academic employment and peer-reviewed publications promises a more innovative approach to knowledge production.

From the standpoint shared by our Martian friend and his unlikely earth-bound allies—Smith, Koyre, Feyerabend, and the local KM guru—cases of research fraud are treated too much as isolated events, deviations from norms that everyone is presumed to follow, rather than symptoms of alternative normative regimes for producing knowledge. These alternatives may make rather counter-normative assumptions about, say, the epistemic superiority of reason to experience (à la Koyre) or risk-seeking to risk-avoidance (à la Feyerabend). But they need not be so exotic. For example, one may simply fake experiments every so often in the direction of the current research trajectory in order to stay within budget. The implied alternative norm of this practice would be that the consumers of scientific research should bear more of the risk entailed by accepting particular knowledge claims, and in that sense, share the burden of experimentation. Indeed, more explicit versions of this norm are familiar from the histories of medicine and technology, which blur, if not erase entirely, the laboratory-based distinction between testing and applying knowledge claims. The "social construction of technology" approach in science studies is founded on just this insight (Bijker, Hughes, and Pinch 1987).

Moreover, this insight was presaged by a venerable strand of British legal thought. Albert Dicey (1981 [1905]), a leading liberal on the Oxford law-faculty in the Edwardian period, famously regarded health and safety legislation as an insult to free citizens, who are entitled to undertake whatever risks they wish, once they have been informed of the relevant consequences. Like today's social constructivists. Dicey held that a wide range of actions may be rationally justifiable on the basis of the available evidence. From that standpoint, health and safety regulations are simply covert attempts by the state to engage in moral instruction, precisely the kind of paternalism that is anti-thetical to liberalism.

A residue of the paternalism that bothered Dicey remains enshrined in John Rawls' (1971) "difference principle," the idea that inequalities are justified only if they benefit the disadvantaged—which is to say, if the rich are better able to help the poor than the poor could themselves, were the difference in wealth transferred to them. However, instead of wealth in sheer material resources, we should think here of a second-order wealth of discretion. Accordingly, human welfare would be increased by the state granting health and safety experts the authority to make decisions on behalf of the entire population than allowing individuals to decide for themselves. This version of the difference principle has been justified by a combination of guild-based public trust assumptions and more a general efficiency consideration, namely, that the experts have compiled the relevant experience that makes further individual experimentation unnecessary. But this is often more presumed than proved. The professionalization of scientific medicine in early twentieth-century United States provides a case in point (Freidson 2001: chap. 8).

When the U.S. Supreme Court decided that the medical profession did not constitute a monopoly operating against the public interest, it pointed to what might be called the proselytizing function of medical doctors in making their services integral to hospitals and local communities, sometimes at significant personal expense and to the apparent benefit of those receiving the services. Epistemologically speaking, however, the expression of positive intent on the part of the doctors and the absence of observable negative consequences (i.e., few lawsuits against doctors) are a far cry from a proper weighing of costs and benefits. Nevertheless, the Supreme Court treated the highly publicized medical campaigns as surrogates—or "signals," as economists would say—for the rigorous accounting on which the difference principle is supposed to be based. Thus, the Supreme Court believed that in granting exclusionary privileges to medical doctors, it was simply ratifying a decision that the American public had already made to entrust their health to scientific medicine.

Routinizing the Genius of Fraud through Information Technology

The history of the American medical profession thus provides a rather sceptical instance of the difference principle in action. Therefore, it would seem to lend support to the liberalized policy toward the knowledge governance proposed by our Martian economist and his new friend, Dicey. Why, then, should individuals *not* be allowed to decide to support or apply a form of knowledge, even if it places them at personal risk? The answer is that no such decision is ever truly private. On the one hand, many "independent" decisions by knowledge consumers can serve to eliminate producers from the market, thereby unwittingly restricting consumers' future choices about the knowledge to which they have access. On the other hand, producers may attempt to gain a greater share of the market by producing goods that more efficiently provide the relevant utilities. In the case of knowledge, the most obvious tendency in that direction comes from automated technology, ranging from the humblest calculator to the most ambitious computerized expert system (Fuller 2002a: chap. 3). Indeed, this tendency has received the philosophical endorsement of the Turing Test, a thought experiment that alleges that a computer could simulate the responses of a human sufficiently well that another human would not be able to distinguish between the two. Our Martian economist would like us to view research fraud as a more fully embodied version of just such an experiment.

From that standpoint, information technology aims to eliminate the "undesirable" side effects of the usual knowledge conveyors (human beings) These include the need to master their specialized modes of communication before gaining access to the relevant knowledge (education), as well as the normally erratic performance levels of humans as both producers and consumers of knowledge (fatigue, emotions, and context sensitivity). At the same time, however, information technology narrows the range of potential interaction between human consumer and computerized expert, thereby reducing the opportunities for the erroneous and otherwise surprising communications that have traditionally served to expand the knowledge bases of humans in both their producer and consumer roles. Moreover, this process can come full circle, if the provision of expert systems is concentrated in a few producers. In that case, consumers may come to depend on a particular information technology to such an extent that they regard the technology's disutilities as an acceptable cost for the utility it provides, much as a student in a vocational or even professional school might treat her education.

The ultimate source of this reduction in the potential for collective knowledge growth is that the access costs to this supposedly public good, knowledge, are now included as part of its production costs. And once knowledge producers themselves—and not some third party, such as the welfare state—are made to bear the cost of enabling consumers to have access to their goods, knowledge inevitably devolves into a private good. This means that what had been a universal need for knowledge is now treated as a market whose boundaries are defined by the number of people who can be served by the same good. This drives production costs down. The role previously played by state-funded education is now assumed by the advertising division of the knowledge producing company, whose costs are then rolled into the price charged to knowledge consumers. As the competition for consumers intensifies, marketing costs rise, perhaps even overtaking production costs. At that point, we enter capitalism of the third order.

If the reader has difficulty imagining this scenario, simply consider the following change in the political economy of knowledge provision. Starting with mandatory public education that is covered by income taxes, we move to a set of alternative software packages, each designed to serve the knowledge needs of a specific range of consumers, based on market research whose costs are recovered as part of the retail price of the packages. In the latter situation, the effort normally involved in understanding difficult material is absorbed into a user-friendly interface, a selling point of which is the minimal effort it demands from consumers. If the market displays little competition, then consumers will blame themselves for whatever remaining difficulties they have in mastering the interface. However, if market competition stiffens, then knowledge producers will be forced to understand their consumers better, so as to instill a mind-set that will make them want to purchase their knowledge products. If this process becomes sufficiently sophisticated, it may start to look like research and education. At that point, the market would acquire strange university-like qualities.

Conclusion

The liberal attitude toward research fraud recalls the romantic image of the free market epitomized in Joseph Schumpeter's (1961 [1912]) heroic entrepreneur, based largely on the career of Henry Ford. The entrepreneur is capable not only of creating a genuinely innovative product, but also of marketing it so as to reap a substantial profit. Schumpeter's original phrase to characterize the entrepreneur's activities, creative destruction, captures the image's appeal. In effect, the entrepreneur manages to catalyse a redistribution of wealth, effort, and desire in society without having to resort to a central planning board with its usual battery of taxes, incentives, and subsidies. Analogously, someone like Galileo could reorient entire fields of knowledge simply because others took up the challenges posed by his work and reaped their own rewards from it. According to the romantic image, neither Ford nor Galileo had to bribe or coerce people into purchasing their goods, even though the true value of those goods had yet to be demonstrated. Rather, the utilities promised by the products encouraged people to take a chance on them. In this way, both producer and consumer end up benefiting from their calculated risks.

Unfortunately, the heroic entrepreneur is at most an ideal type, not the statistical norm of economic action. After all, entrepreneurs turn out to be heroic to the extent that they make it more difficult, if not impossible (à la Bill Gates), for competitors to operate in the newly reconfigured market. In this respect, a monopolistic market and a paradigmatic science are two versions of the same post-entrepreneurial (post-revolutionary) situation. In markets with many successful entrepreneurs, their "success" tends to be rather more modest, uneven, and temporary—a situation not unlike the multiple schools that constitute the humanities and social sciences at any given moment. Moreover, the ratio of successful to unsuccessful entrepreneurs remains very low.

Nevertheless, even when portrayed in a more realistic light, the image of the entrepreneur remains attractive for reasons related to Kitch's view of knowledge as self-protective. Successful entrepreneurs often seem to come from unexpected backgrounds that enable them to see already existing knowledge in a new light, and consequently make more of it than the original knowledge producers themselves did. Yet, even granting this point, saying that knowledge tends to escape from its original bearers is a far cry from saving that knowledge tends to be made universally available. The former is simply a long-winded acknowledgement that communication exists; the latter implies that all communication is mass communication. *However, the fact that knowledge is not a private good does not mean that it must therefore be a public good.* Kilch's vision comes dangerously close to this non sequitur, which overlooks the importance of institutions in the construction of knowledge as a public good. In particular, the university's intergenerational responsibilities regularize the creative destruction of the social capital in a way that no real market is ever likely to do.

I have argued that there is nothing inherently fugitive about knowledge that renders it ungovernable. Rather, knowledge began to acquire this character in the leisured pursuits of ancient Athens, which also doubled as a means of demonstrating civic privilege. In contrast, it was only with the establishment of the first universities that knowledge came to resemble what is now-called a public good, mainly because the institution was designed so as to create and destroy societal advantage simultaneously through, respectively, research and education. The identity of knowledge as a public good strengthened with the consolidation of the nation-state in the nineteenth century, culminating in the welfare states of the second half of the twentieth century. In this period, universities came to acquire their current prominence as knowledge production sites. However, as evidenced in Kitch's legal vision, the ongoing devolution of the welfare state is eroding our intuitions of knowledge as a public good. Indeed, knowledge managers today portray universities *at their best* as less creative destroyers than value-adding accumulators, who are sometimes in direct competition with private knowledge producing firms (Brown and Duguid 2000). In this changing environment, perhaps the regime of knowledge governance best suited to universities is to become producers of what the economist Michael Perelman (1991) calls *metapublic goods.* The university would then become a corporate trust buster dedicated to the prevention and subversion of all attempts to turn knowledge into a form of social capital, even if it can no longer ensure knowledge's status as fully fledged public good (Fuller 2002a: chap. 1).

Acknowledgments

I would like to thank Nico Stehr for his indulgence, the references to his own work, as well as those of Gernot Böhme, Wolfgang van den Daele, and especially Edmund Kitch. Also, thanks to Peter Ploeger for insight into the German context in which universities struggle for both autonomy and relevance.

References

Anderson, Benedict (1983) *Imagined Communities: Reflections on the origin and spread of nationalism*. London: Verso.

Bell, Daniel (1966) *The Reforming of General Education: The Columbia College experience in its national setting*. Garden City, N.Y.: Doubleday.

Bijker, Wiebe E., Thompson P. Hughes, and Trevor J. Pinch, eds. (1987) *The Social Construction of Technological Systems: New directions in sociology and history of technology*. Cambridge: MIT Press.

Bloor, David (1983) *Wittgenstein and the Social Theory of Knowledge*. Oxford: Blackwell.

Böhm-Bawerk, Eugen von (1959 [1884]) *Capital and Interest: History and Critique of Interest Theories*. George D. Huncke and Hans F. Sennholz (trans.). South Holland, Ill.: Libertarian Press.

Bourdieu, Pierre (1975) "The specificity of the scientific field and the social conditions of the progress of reason," *Social Science Information* 14(6): 19–47.

Brown, John Seeley, and Paul Duguid (2000) *The Social Life of Information*. Cambridge, Mass.: Harvard Business School Press.

Collins, Randall (1998) *The Sociology of Philosophies: A global theory of intellectual change*. Cambridge: Harvard University Press.

Conant, James Bryant (1970) *My Several Lives: Memoirs of a social inventor*. New York: Harper and Row.

Daele, Wolfgang van den, Alfred Pühler, and Herbert Sukopp (1997) *Transgenic Herbicide-Resistant Crops: A participatory technology assessment*. Discussion paper FS II 97–302. Wissenschaftszentrum Berlin für Sozialforschung.

Dicey, Albert (1981 [1905]) *Lectures on the Relation Between Law and Public Opinion in England During the Nineteenth Century*. Oxford: Clarendon Press.

Duhem, Pierre (1991) *German Science*. La Salle, Ill: Open Court Press.

Faundez, Julio (1994) *Affirmative Action: International perspectives*. Geneva: International Labour Organization.

Feyerabend, Paul (1975) *Against Method*. London: Verso.

Freidson, Eliot (2001) *Professionalism: The third logic*. Chicago: University of Chicago Press.

Fukuyama, Francis (1992) *The End of History and the Last Man*. New York: Free Press.

Fuller, Steve (1993) *Philosophy of Science and Its Discontents*. 2nd edition. New York: Guilford Press.

—— (1993) *Philosophy, Rhetoric and the End of Knowledge*. Madison, Wisc: University of Wisconsin Press.

—— (2000a) *The Governance of Science*. Milton Keynes, U.K.: Open University Press.

—— (2000b) *Thomas Kuhn: A Philosophical History for Our Times*. Chicago: University of Chicago Press.

—— (2001) "The Darwinian Left: A rhetoric of realism or reaction?" *POROI* 1(1). http://inpress.lib.uiowa.edu/poroi/vo12001/fuller2001.html.

—— (2002a) *Knowledge Management Foundations*. Woburn, Mass: Butterworth-Heinemann.

—— (2002b) "The University: A social technology for producing universal knowledge," *Technology in Society*. . . .

Gates, Henry Louis (1988) *The Signifying Monkey*. Oxford: Oxford University Press.

Gibbons, Michael, Camille Limoges, Helga Nowotny, Simon Schwartzman, Peter Scott, and Martin Trow (1994) *The New Production of Knowledge: The dynamics of science and research in contemporary societies*. London: Sage Publications.

Godelier, Maurice (1986) *The Mental and the Material*. London: Verso.

Gregory, C.A. (1994) "Exchange and Reciprocity," in *Companion Encyclopedia of Anthropology*. T. Ingold (ed.). London: Routledge. 911–39.

Grundmann, Reiner and Nico Stehr (2001) "Why is Werner Sombart not part of the core of classical sociology?" *Journal of Classical Sociology* 1: 257–87.

Guston, David (2000) *Between Politics and Science*. Cambridge, U.K.: Cambridge University Press.

Hagstrom, Warren (1965) *The Scientific Community*. New York: Basic Books.

Hirsch, Fred (1977) *The Social Limits to Growth*. London: Routledge and Kegan Paul.

Keller, Evelyn Fox (1983) *A Feeling for the Organism*. New York: Freeman.

Kitch, Edmund (1980) "The law and the economics of rights in valuable information," *The Journal of Legal Studies* 9: 683–723.

Koyre, Alexandre (1978 [1939]) *Galileo Studies*. J. Mepham (trans.). Atlantic Highlands, N.J.: Humanities Press.

Krause, Elliott (1996) *The Death of the Guilds: Professions, States, and the Advance of Capitalism*. New Haven, Conn.: Yale University Press.

Lutz, Mark (1999) *Economics for the Common Good*. London: Routledge.

Merton, Robert (1973) *The Sociology of Science: Theoretical and empirical investigations*. Norman W. Stored (ed.). Chicago: University of Chicago Press.

Moore, Barrington (1963) *The Social Origins of Dictatorship and Democracy*. Boston: Beacon Press.

Nicholson, Linda, ed. (1994) *Feminist Contentions*. London: Routledge.

Noble, Douglas (1991) *The Classroom Arsenal*. London: Falmer Press.

Olesko, Kathryn (1993) "Tacit Knowledge and School Formation," in *Research Schools*, G. Geison and F. Holmes (eds.) *Osiris*, New Series, Volume 8. Chicago: University of Chicago Press. 16–29.

Perelman, Michael (1991) *Information, Social Relations, and the Economics of High Technology*. New York: St Martin's Press.

Penslar, Robin, ed. (1995) *Research Ethics: Cases and Materials*. Bloomington, Ind.: Indiana University Press.

Pigou, Arthur (2000 [1920]). *The Economics of Welfare*. New Brunswick, N.J.: Transaction.

Price, Derek de Solla (1978) "Toward a Model of Science Indicators, in *Toward a Metric of Science: The advent of science indicators*. Yehuda Elkana, Joshua Lederberg, Robert K. Merton, Arnold Thackray, and Harriet Zuckerman (eds.). New York: Wiley-Interscience. 69–96.

Rawls, John (1971) *A Theory of Justice*. Cambridge: Harvard University Press.

Samuelson, Paul (1969) "Pure Theory of Public Expenditures and Taxation," in *Public Economics*, J. Margolis and H. Guitlon (eds.). London: Macmillan. 98–123.

Schaefer, Wolf, ed. (1984) *Finalization in Science*. Dordrecht, Neth.: Kluwer Academic Press.

Schumpeter, Joseph (1950 [1942]) *Capitalism, Socialism, and Democracy*. New York: Harper and Row.

—— (1961 [1912]) *The Theory of Economic Development*. Chicago: Galaxy Books.

Stehr, Nico (1994) *Knowledge Societies*. London: Sage Publications.

Woodmansee, Martha (1984) "The genius and the copyright," *Eighteenth Century Studies* 17: 425–48.

Part II

Major Social Institutions and Knowledge Politics

Introduction

Martin Schulte

Knowledge production has been the subject of vigorous debates in the social sciences, at least since Karl Mannheim's and Max Scheler's surveys of the social bases of knowledge at the beginning of the last century (Scheler 1960a and 1960b: Maasen 1999). Particular attention, however, was devoted to the topic again only in the early 1990s, initially in science studies and then in connection with the transition from a society of labor and property (industrial society) to the knowledge society (Stehr 1991a, 1991b, 1994; 2001, 2002). In both cases, the debates were highly contested (for example, the finalization debate and the so-called science wars) and resonated with discussions among public and policy makers. More recently, these issues have also been taken up by legal scholars and the legal system (Bora 1999; Engel, Halfmann, and Schulte 2002; Schulte 2003).

At the present time, the question of knowledge production is increasingly accompanied and to some extent displaced by discussions of the problems of knowledge policies in modern societies investigating the prospects of the social regulation of the rapidly growing body of new knowledge, and therefore the supervision as well as control of knowledge. The social system of law and jurisprudence is challenged and becomes involved, raising an admonitory voice to deal more actively with the question of legal control of the production and use of knowledge (Grundmann and Stehr 1999; Stehr 2001b).

Against this emerging background, the contribution by Werner Rammert in this section examines the sociological bases of modern knowledge policy, in particular the increasing importance of non-explicit or tacit knowledge. The contributions by William Leiss and Jeffrey Klein in turn focus on practical knowledge policy issues, that is, "genetics, nanotechnology, robotics," and "bioinformatics."

In Rammert's opinion, the production of knowledge is more and more turning into an economic enterprise in which universities and research institutes take about an equal share. Knowledge, however, is considered to be more of an ability (competence) to do something rather than a commodity that can be sold, bought, and transported. But the decisive factor according to Rammert is the differentiation of knowledge production. Rammert therefore differentiates between segmental differentiation ("the same knowledge is dispersed over many places without any mechanisms of coordination, like exchange or central collection"), hierarchal differentiation ("people start to collect the knowledge from the dispersed places, and some organizations begin to systematize and centralize the relevant knowledge") and functional differentiation ("spheres of actions and action-orientations are separated horizontally from each other ... by generating a self-referential orientation following only their own code of orientation purifying it from other influences").

Rammert also advocates another type of social differentiation, that is, fragmentary differentiation, which he considers to be the emerging and dominant form of differentiation in science in the future. From his point of view, this type of social differentiation "shares the feature to combine different types of knowledge with the segmental one, but it differs from it under the aspect that its mix of heterogeneous knowledge consists of fragments of a once systematized and functionally specialized knowledge. Elements of all kinds of knowledge are recombined. The fragmental differentiation differs from the functional one radically under the aspect that the purified separation is given up in favor of heterogeneity and reflexivity." For Rammert, this form of social differentiation opens up entirely new perspectives: The network becomes a novel mechanism of coordination, apart from hierarchy and market; idea-innovation-networks become principal agents of knowledge production and innovation; innovation as such is mainly based on heterogeneous and interactive networks; the multidisciplinary, or pluridisciplinary, knowledge production of "mode 1" has to be extended by the interdisciplinary, and transdisciplinary, knowledge production of mode 2 (see Fuller 2000: 79–83; Nowotny, Scott, and Gibbons 2001: 88–95).

While Rammert is discussing the sociological foundations of modern knowledge policy, the contributions by Leiss and by Klein mainly focus on the topical fields of the implementation of such policies in light of concrete developments in different fields of knowledge production.

Leiss is concerned, from a legal perspective, with two fields of current interest: on the one hand, how risk is to be handled in the constitutional state, and on the other, how knowledge, the lack of knowledge, and uncertain knowledge are to be handled under the conditions set by an increasingly global society.

While the risk-problematic has already been discussed in sociology for some time, and a "sociology of risk" has been established (e.g., Japp 2000), the career of "risk" as a legal concept has become familiar only the last decade. It was Udo di Fabio (1994) who introduced the discussion of risk into legal scholarship with his discussion of "Decisions on issues of risk in the modern constitutional state." In the meantime, however, a juridico-conceptual approach has been developed that seeks to steer issues on how knowledge, lack of knowledge, and uncertain knowledge are to be handled in the law. Against this background, it is possible to subject the risks examined by Leiss (moral risks, catastrophic risks) to an assessment from the point of view of the social institution of law and legal science in modern society.

In this connection, the legal specialist distinguishes, above all, between danger and risk (see also Luhmann 1993). Danger refers to a certain probability, namely, that a substantial adverse impact on a good that is protected by a special code (e.g., environmental law) will take place. The greater the impact of the risk feared, the less exacting are the requirements to be met in order to offset its occurrence. Risk, however, means the possibility of the occurrence of a significantly adverse impact that is protected by a special code, except insofar as it appears to be practically excluded. The concept of risk therefore aims at those possible occurrences, deemed disadvantageous, that fall below the threshold of danger. According to the precautionary principle, risks to the environment or to human beings shall be excluded or at any rate minimized to the extent possible, in particular, by means of forward-looking planning and suitable technical precautions. Precautionary measures, where risk is concerned, are to be sure, required only within the limits of what is technically possible; they stand, in other words, under the reservation that there be an appropriate relation between the effort expended and the outcome that can reasonably be expected.

The concepts of risk serve to distinguish risk aversion from precautions taken against risk. Insofar as this concept is concerned, and given the present state of science and technology, it is practically ruled out that life, health, or material goods will suffer damage or harm. Beyond this "threshold of practical reason" (as defined by the German Federal Constitutional Court for example), the remaining uncertainties have their origins in the limits of our cognitive powers.

The legal system does not require risk precautions in this instance, for such uncertainties are inescapable and count as burdens that are socially tolerable and therefore must be borne by all citizens. There is no such thing as absolute certainty—not, at any rate, so long as one is not willing to forego technology as we know it. From the known risks, the potential risks are distinguished by the fact that knowledge informed by experience will be lacking for as long as we remain at the limits of our cognitive powers. To be sure, a potential risk can turn into an actual risk or, indeed, into a danger if our cognitive powers—on the basis of scientific or technical advances—are extended beyond their present limits.

How to most effectively meet and incorporate into the life world these special and, above all, global risks of which Leiss is speaking is a question that reaches far beyond existing legal codes to a vision that posits the establishment and the limits of a system of world law. Doubtlessly, the global interconnections of all the functional systems of society are beyond dispute. In a very special way, this is true for the economy and for science, which under modern conditions have to be thought of in globally societal terms. Not without reason, however, one needs to be cognizant of distinctive features of the legal system and the political system. Thus, in spite of all word-wide interconnections of the legal system, major and serious differences in the legal systems of the various regions of the world are unmistakable. Whether these differences can in fact be overcome for the sake of a kind of "world law"—quite apart from the question of whether one deems this desirable— appears highly questionable. As a result, one will also have to be cautious in evaluating efforts to monitor, with global standard and codes in mind, the creation, use, and distribution of new forms of knowledge associated with the kinds of risks discussed.

The risks of biotechnology, in particular bioinformatics, are investigated by Klein. From his point of view, scientific advances encouraged by the development of technology, and technological development, in turn, is promoted and motivated by the interest of industry. Of course, the state, governments, and various administrative corporate actors will try to maintain their central position as regulatory authorities, but this probably will be unsuccessful in the long term. It is progress alone—an almost divine value of its own—that pushes the revolution in biotechnology forward.

The elitist circles of the society, rational materialists and the profoundly utilitarian, are hard to control. Even science and the universities will not achieve this purpose, because any attempt to regulate the use of knowledge is bound to be undermined by the industry and the technologists. The attempt at concealing scientific controversy is just as shocking an alternative. According to Klein, the only way out would be to make as many people as possible participate both in the production and use of knowledge. The promises of egalitarianism, including wide public participation—that is the model he suggests for the future of knowledge policies in modern societies.

References

Bora, Alfons (1999) *Rechtliches Risikomanagement.* Form. Funktion und Leistungs-fähigkeit des Rechts in der Risikogesellschaft. Berlin: Duncker and Humblot.

Di Fabio, Udo (1994) *Risikoentscheidungen im Rechtsstaat.* Zum Wandel der Dogmatik im öffentlichen Recht, insbesondere am Beispiel der Arzneimittel-überwachung. Tübingen: J.C.B. Mohr (Paul Siebeck).

Engel, Christoph, Jost Halfmann, Schulte, Martin (2002). *Wissen, Nichtwissen, Unsicheres Wissen.* Baden-Baden: Nomos.

Fuller, Steve (2000) *The Governance of Science. Ideology and the future of the open society.* Buckingham: Open University Press.

Grundmann, Reiner and Nico Stehr, November 23 (1999). "Wissenspolitik ist Macht," *Frankfurter Allgemeine Zeitung* (273): 54.

Japp, Klaus (2000) *Risiko.* Bielefeld: Transkript

Luhmann, Niklas (1993) *Risk: A sociological theory.* New York: Aldine de Gruyter.

Maasen, Sabine (1999) *Wissenssoziologie.* Bielefeld: Transkript.

Nowotny, Helga, Peter Scott and Michael Gibbons (2001) *Re-Thinking Science: Knowledge and the public in an age of uncertainty.* Oxford: Polity Press.

Scheler, Max (1960a) "Probleme einer Soziologie des Wissens," in *Die Wissens-formen und die Gesellschaft,* Maria Scheler (ed.). Bern: Francke.

—— (1960b) "Erkenntnis und Arbeit," in *Die Wissensformen und die Gesellschaft,* Maria Scheler (ed.). Bern: Francke.

Schulte, Martin (2003) "Wissensgenerierung in Recht und Rechtswissen-schaft—Selbstbeschreibung und Fremdbeschreibung des Rechtssysterms," in *Szenarien der Wissensgesellschaft,* Armin Nassehi (ed.). Frankfurt am Main: Suhrkamp.

Stehr, Nico (1991a) "The power of scientific knowledge—and its limits," *Canadian Review of Sociology and Anthropology* 29: 460–82.

—— (1991b) *Praktische Erkenntnis,* Frankfurt am Main: Suhrkamp.

—— (1994) *Knowledge Societies.* London: Sage Publications.

—— (2001a) *The Fragility of Modern Societies.* London: Sage Publications.

—— (2001b) "Moderne Wissensgesellschaften," *Aus Politik und Zeitgeschehen* B36: 7–13.

—— (2002) *Knowledge and Economic Conduct: The social foundations of the modern economy.* Toronto: University of Toronto Press.

4

The Rising Relevance of Non-Explicit Knowledge under a New Regime of Knowledge Production

Werner Rammert

Scientific and technological knowledge is going to become the key factor in the change of economies and in the evolution of society. The production, distribution, and use of knowledge determines the competitive advantages of enterprises (Machlup 1962). It increases the innovative capacities of economies. It strengthens the power position of nation-states in the world-system. As the production of knowledge is dispersed over many places, as it is split into the use of explicit and non-explicit knowledge, and as it is divided between different institutional actors, new means and mechanisms of coordination are demanded beyond market and hierarchy. "Interactive networks of innovation" (Lundvall 1993) are constructed in order to gain access to local knowledge and share profit and risk of its global use. "Knowledge management" is created as a new field of business in order to raise the efficiency of knowledge work and to keep up with the pace of innovation. "Knowledge politics" is coming up as a new field of policy. Its aim is to establish the adequate institutional infrastructure that assures and accelerates the growth of knowledge production. If one shifts the focus from the division of labor that predominated in industrial society to the division of knowledge (Hayek 1945; Helmstädter 2000) that takes place actually, one can observe and perhaps explain the emergence of a new regime of governance. I shall call it the *regime of distributed knowledge production*.

There are two features of the actual mode of knowledge production: On the one side, the knowledge production has turned into a business.

Universities and research institutes are transforming themselves into patent-holding and knowledge-selling enterprises. Firms and companies engage themselves in the business of knowledge creation and knowledge management. Scientific institutions, state agencies, and industrial R&D laboratories are knitted together to build either national innovation systems (Nelson 1993) or networks of innovation that operate between heterogeneous actors and on an international level (Freeman 1992; Powell, Koput, and Smith-Doerr 1996). Their activities are coordinated more and more under the imperatives of economic innovation and national wealth (Foray and Freeman 1993). The first aim of this paper is to give a short outline of the institutional changes that are leading to a new regime of distributed knowledge production.

On the other side, limitations of the governance of knowledge become visible. They are rooted mainly in two paradoxical processes. First, the heterogeneity of the enrolled actors—scientists and managers, politicians and administrators, venture capitalists and ecological activists—and the diversity of their perspectives cause problems of a successful concertation (Rammert 2000) that does not level out the creative differences between disciplines or institutional rationality standards, and that does not destroy the complementary competences of functionally specialized actors. Second, the specificity of knowledge to be an intangible asset and an incompletely explicable set of competences sets limits on the complete control and commercialization of knowledge. The second aim of this paper is to explore the *role of non-explicit knowledge* in the process of making a growing part of the knowledge more explicit.

The innovativeness and the economic performance of the rising science-based "knowledge societies" (Böhme and Stehr 1986) are fundamentally determined by the ways they cope with both problems, the institutional problem of coordinated distributedness between heterogeneous actors and the epistemological paradox of the explication of the non-explicit knowledge.

The Rise of a New Regime of Governance: Distributed Knowledge Production under Conditions of Fragmentation

Knowledge has to be seen more as a competence to do something than a compact good that one can transport and store (Rammert et al. 1998: 37; Stehr 2000: 81). A scientific paper or a written patent claim are, however, packaged pieces of information; but they can only be converted into knowledge if people or organizations know how to use them. The two aspects, the incorporated "known" and the enacted "knowing" (Dewey and Bentley 1949), belong together and should not be analyzed separately.

If we accept this pragmatic definition of knowledge, then it follows that it makes a difference how the system of knowledge production is institutionalized in society. Following a widely recognized theory of societal differentiation (Luhmann 1977; Munch 1990), we can distinguish between three types of differentiation: the segmental, the hierarchical, and the functional. Let us assume that there exists a close interrelationship (*Wahlverwandtschaft*) between a type of social differentiation and a regime of knowledge production. It is not a strong causal relation, but it is supposed that particular patterns of dominant social structures favor certain means of coordination and suggest specific institutional answers. These institutional answers show a great variety in the beginning. But after periods of transformation, the trial-and-error process of learning is closed. A new regime of governance is established that consists of a selected set of institutions and that is characterized by the dominance of particular means of coordination. Dominance does not mean that the other types and means are diminishing or even disappearing. The rise of a new regime only indicates that one type of differentiation achieves the leading role in shaping society whereas the others only loose their centrality, but gain new importance for particular social units.

The term governance of knowledge should as well not be misunderstood: Knowledge production, especially scientific knowledge production or even technological innovation, cannot be guided or regulated like car production or conventional industrial production systems. The uncertainties of innovation are incomparably higher than those of product markets. The heterogeneity of actors and the diversity of their orientations are so complex that no principle agent can appropriate the necessary information and control the actions of all of the agents from different fields. That is why the state is no longer seen as the central authority; even state regulation must be divided between the government and private or semi-private agencies. New forms of divided systems of governance evolve, like the regime of "corporate governance" (Hollingsworth, Schmitter, and Streeck 1994) that enrols the most important collective actors and excludes the rest, or the "distributed governance" by "policy networks" (Marin and Mayntz 1991) where all relevant actors of a political field participate in the shaping of the output.

In the following section I shall roughly characterize the three widely acknowledged types of societal differentiation and relate them to idealized regimes of knowledge production.

Segmental differentiation is characterized by the split into many parts of the same kind and of the same status. Families, clans, and tribes are the resulting social units that dominated in archaic societies. Segmentation is a kind of homogeneous division. Clan relations and myths are the weak means of coordination between the homogeneous and autonomous parts. Under conditions of segmental differentiation, the same knowledge is dispersed over many places without any mechanism of coordination, like exchange or central collection. Every tribe and every settlement controls its own local knowledge production. Internally, special roles like artisan or medicine man are separated; but one can find the same stock of knowledge in every social unit. A regime of *local* and *dispersed* knowledge production emerged that affiliates with the segmental type of social differentiation.

Hierarchal differentiation divides the whole into parts of a different kind and with different status. It establishes a system of vertical distinction. Historical examples are stratifications between high and low classes or between center and periphery. Empires based on military forces and monotheistic religions with a missionary impetus are the strong means of coordination between the distinguished ranks and between the various places. Under conditions of hierarchical differentiation, people start to collect the knowledge from the dispersed places, and some organizations begin to systematize and centralize the relevant knowledge. Such people were called "Mechanici," technicians and scholars. The monastery, the early universities, and the city guilds were the early places of knowledge collection and exchange. The church and, later on, the state became the principal agents who controlled the knowledge production and diffusion. A regime of *universal* and *centered* knowledge production came up that co-evolves with the hierarchical type of social differentiation.

Modern society differs from traditional society by the predominance of *functional differentiation*. It splits society into complementary parts that are organized around different functions, but which have the same status. Spheres of actions and action-orientations are separated horizontally from each other, like the economic, the political and the scientific subsystems. They differ under the aspect of a specialized contribution to the reproduction of society. They increased their efficiency by generating a self-referential orientation following only their own code of orientation purifying it from other influences. They achieved relative autonomy from outside interventions by establishing a system of self-organization. As the functions are indispensable and cannot be substituted by another subsystem, all functionally specialized systems are equally important. A system of horizontal division outplays the vertical system of stratification. The production and use of economic, governmental, and scientific knowledge is institutionalized in these specialized spheres of society. Society loses its center and central principle of organization. Markets, mobilization of resources by state or political movements, and discourses (scientific and public) advance to equivalent means of coordination. Scientific knowledge production, for example, gains high institutional autonomy and self-governance; but in order to exploit it for the sake of economic innovation or military power enforcement, markets of patents and licenses and big organizations were recruited to coordinate the separated, but complementary innovative activities. A regime of *complementary* and *specialized* knowledge production belongs to the functional type of social differentiation.

But the consequences of this functional differentiation lead to unintended problems of synchronization and adjustments that some analysts call "reflexive modernization" (Beck, Giddens, and Lash 1994). Other analysts conceive these changes as formations of secondary subsystems that operate to cope with the consequences of the primary subsystems. Some others diagnose them as processes of "de-differentiation" or "heterogeneization" (Weingart 1983; Tyrell 1978; Knorr Cetina 1992). I do not want to continue this debate on the basis of three types of differentiation, but I want to give it a fresh impetus I claim the logical possibility (Schimank 1996: 151) and the empirical validity of a fourth type of social differentiation: fragmental.

The *fragmental differentiation* splits a heterogeneous whole into parts that are of the same kind, but of a different status or on a different level. Regional innovation networks, for example, always include nearly the same mixture of elements, like political, economic and cultural actors and institutions. But some networks are setting the bench marks, like the Silicon Valley network of microelectronics and software industry, or the Baden-Württemberg network of mechanical engineering and car production, others are imitators and followers in the global competition. The differentiation of scientific disciplines is rooted in well-defined and theory-bound fields of research. Each enjoys the highly reputed status as ascertained knowledge that we call scientific truth. But when the number of mission- and issue-oriented scientific research projects increases, the new type of fragmented knowledge production is coming up that is rooted in the same combinations of heterogeneous relevant knowledge. The quality is no longer reviewed by the peers of one discipline, but by heterogeneous expert groups and epistemic cultures.

Fragmentation is a kind of division that combines heterogeneous elements. It shares the feature to combine different types of knowledge with the segmental one, but it differs from it under the aspect that its mix of heterogeneous knowledge consists of fragments of a once systematized and functionally specialized knowledge. Elements of all kinds of knowledge are recombined. The fragmental differentiation differs from the functional one radically under the aspect that the purified separation is given up in favor of heterogeneity and reflexivity. Functionally specialized institutions and purified scientific disciplines remain important factors on the back stage of fragmented society, but they are losing their privilege to act on the front stage where network forms of organization and transdisciplinary epistemic and export cultures take over the prominent roles. The *regime of heterogeneously distributed knowledge production* rises in close relation to the type of fragmental differentiation.

If we have once accepted the new idea that a fourth type of social differentiation can be deduced theoretically, then many empirical studies in the fields of neo-institutionalism and of science and technology

Table 4.1. Types of Societal Differentiation and Regimes of Knowledge Production.

Societal Differentiation	Distinctive Features	Means of Coordination	Regime of Knowledge Production
segmental	dispersed, homogeneous division	clans myths	local and dispersed
hierarchal	concentrated, vertical division	empire religion	universal and centered
functional	separated, horizontal division	market media science	complementary and specialized
fragmental	combined, heterogeneous division	networks epistemic cultures	heterogeneous and distributed

studies can be seen under a new perspective. They describe the emergence of this new type of social differentiation and the development of the new regime of heterogeneously distributed knowledge production.

- The "network form of organization" was discovered as a particular mechanism of coordination in addition to hierarchy and market (Powell 1990).
- "Idea-Innovation networks" are described as the principle agents of knowledge production and innovation (Hage and Hollingsworth 2000).
- A post-Schumpeterian mode of "reflexive innovation" based on heterogeneous and interactive networks of innovation is distinguished from two modern ones based on Schumpeterian inventive entrepreneurs and on big slate, big science, or big business bureaucracies. This network form of organization is identified as the adequate institutional mechanism to synchronize the different tempi of innovation between the scientific, economic, and political system (Rammert 1999a; 2000).
- A "mode 2" of a transdisciplinary and reflexive production of scientific knowledge was claimed to develop in addition to the mode 1 of normal disciplinary knowledge production. One of its characteristics is the integration of political and moral knowledge into the scientific knowledge production (Gibbons et al. 1994; Nowotny et al. 2001).
- Hybrid "actor-networks" are described as the relevant unit and socio-technical association whereby scientific facts and technological artifacts are constructed. The borderlines between human and nonhuman entities and between the humanities, social sciences and technosciences were deconstructed (Latour 1994; Gallon 1993).
- "Epistemic cultures" were said to shape the scientific knowledge production, not a set of purified disciplinary knowledge. An epistemic culture encompasses scientific and non-scientific practices and objects (Knorr Cetina 1999).
- "Distributed cognition" was the phenomenon that Ed Hutchins discovered when he observed knowledge production for the sake of ship navigation. It was not the functionally specialized or centrally organized knowledge that was relevant during the breakdown of the automatic navigation system. It was a kind of natural distributed process that consists of accommodating practices (Hutchins 1996).

One may doubt whether all these empirical phenomena and these new categories of description really designate the emergence of a qualitatively new type of knowledge production (e.g., Weingart 1997). But one cannot deny the current changes of practices, discourses, and institutional rearrangement in the process of knowledge production. Under which aspects do they differ from the standard model of knowledge production in functionally differentiated modern societies, and especially, what are the implications concerning the governance of knowledge?

The specialization of disciplinary and subdisciplinary knowledge production has multiplied the research fields with the consequence that aside from new specialties and interdisciplinary cooperation new fields of technoscience emerge under the aspect of cognitive integration, only loosely coupled and remaining heterogeneous; like robotics or nanotechnology, for example, they are more strongly coupled by sharing certain practices of borderline activities and a "pidgin" type of communication (Galison 1996; Meister 2002).

The complexity of the research objects is highly increased by the expansion of the aspects one is interested in, for example, sustainable technical systems of energy production instead of efficient energy production, or by the enlargement of the perspective, for example, from narrow weather forecasting to complex climate research (Stehr and von Storch 1999).

The integration of hardware and software products into situations of work and communication enforces the rise of research fields that develop transdisciplinary models and methods to grasp the hybridity of the research object, for example, in high-tech workplace studies (Button 1993, Star 1996) or in socionics (Malsch 1998; 2001).

It is not a really new phenomenon, especially in the engineering sciences, that the objects of research are *complex* under the aspect of elements and relations, *heterogeneous* under the aspect of included disciplinary perspectives, and *hybrid* under the aspect of kinds of focused objects. But it seems to me that the knowledge production that crosses the traditional disciplinary borders actually gains a predominant role, like in neurological computing, nanotechnology, genetic engineering, or robotics. The knowledge situation can better be described to be in the state of a *loosely coupled distributedness* than in the state of a functional or even a hierarchical integratedness. It is far beyond the phase of finalization (Böhme, van den Daele, and Weingart 1976). It looks more like a puzzle of disciplinary knowledge fragments.

What are the consequence of this fragmental distributedness for knowledge politics? The role of disciplinary knowledge production has to be maintained because one needs the inputs of the disciplinary stock of knowledge and repertoire of methods. But the disciplines lose their overall structuring force; new forms of co-evolution, shared testbeds and meta-languages of communication (Galison 1996) are developing and substituting the strict control of disciplinary theory. Knowledge policy has to foster the development of innovation networks. It should control the balances within and the access to them by heterogeneous, even oppositional actors. It should encourage the participation on each level, for example, by interactive workshops between experts and citizens, by organized mediation between industrial, political, and scientific groups, or by establishing networks of cooperation or platforms of communication that enhance the institutional learning between science, economy, and polity.

The acceleration of the tempo of knowledge production has limited the coordinative capacities of the linear-sequential model of innovation. The standard model produced certainty by connecting the stages of inception, invention, production, and diffusion consecutively. The different activities from the idea-creation to the marketing of knowledge are now organized in several functionally specialized arenas (Hage and Hollingsworth 2000), and they are performed concurrently as in parallel computing processes, not one after the other.

The acceleration is closely connected with the differentiation of spheres and arenas. This interrelationship causes problems of synchronization between the different tempi (Rammert 2000). Discontinuity instead of continuity dominates. Continuity can only be acquired by permanent and parallel interactions between the diverse actors and agencies on several levels.

What are the consequences of this discontinuity and simultaneity for knowledge policy? Different tempi of development have to be recognized in the different institutional fields in order to keep open the time-horizons, for example, a fifty-years cycle for the reorganization of disciplinary knowledge, a fifteen-years cycle for transdisciplinary research fields, a five-year cycle for domains of product innovation, or a one-year cycle for marketing-knowledge. A good governance would avoid rigid time schedules and would not force all under a unified time regulation that equalizes what should be different. The governance of knowledge means under these conditions "making happening, not making strictly." The role of knowledge policy should be to maintain the space for different time-horizons and to construct continuity by permanent and parallel interactions on several levels by the means of mediation, not by the techniques of dictating.

The multiplication of agents who take part in the knowledge production leads to new role requirements and to fragmental units that reproduce the same set of competences at every place, for example, universities supplement their traditional research and teaching competences with new management, fund-raising and start-up competences, whereas business enterprises develop scientific education and research competences as well as their management and venture competences.

The pluralization of actors' perspectives and of institutional contexts raises severe problems of coordination and quality control. Under the conditions of functional differentiation, the interfaces between science, industry, and government could be managed with the help of highly standardized procedures and hierarchically organized advisory groups. But under the conditions of distributed and fragmented knowledge production, new forms of mediation and interactive networking are emerging, for example, mediation has been established as a new profession; consultance concerning knowledge production, knowledge management, and knowledge politics have grown into a successful industry; the task of initiating and moderating networks of knowledge production has become a quite visible concern of state agencies.

What are the consequences of this multiplication of agents and pluralization of perspectives for knowledge politics? As the body of knowledge is bigger than at any time before, as the frames, places and the portals of knowing are growing like "mille plateaux" (Deleuze and Guattari 1992), the established mechanisms of integration, like curricular education, public opinion, and mass communication, are not sufficient. Governance of knowledge that takes place under these conditions of proliferous growth of knowledge production and a fragmented public is challenged to aim at both the maintenance of the creative diversity of actors, opinions, and perspectives as well as the institutionalization of codes, cultural models, and procedures that enable processes of collective learning (Rammert 2002).

The Limits of Explication and the Rising Relevance of Non-Explicit Knowledge

All societies construct themselves in distinguishing some practices and parts of knowledge and giving them an exclusive meaning. Premodern societies used myths, rituals, and monuments to express their systems of relevances. Modern societies can be characterized by making more and more explicit the tacit and traditional knowledge. They privilege explicit knowledge in comparison to all types of non-explicit knowledge. They are more interested in the production of new than in the circulation of old knowledge.

What are the most eminent fields in modern society where knowledge is made explicit? It is the field of science where technological experiences and practical knowledge are transformed into scientific laws. It is the field of economy where commercial intuition and rules of thumb are cast into the calculus of profit and loss accounts. It is the field of law and legislation where moral rules and political priorities are made explicit in the process of codification. Finally, it is the field of complex organizations where a written and formalized form is given to the tasks and duties of its members. Rational science, capitalist economy, positive law, and complex bureaucracy are the resultant modern institutions that can be characterized by their high grade of explicitness.

If we accept the diagnosis that we are approaching a type of society that is based on the axis of knowledge production, then the modern imperative of making knowledge explicit is expected to be amplified and intensified:

- Scientific knowledge penetrates all fields of praxis and has reached the status of the first productive force.
- The economical production, appropriation, and evaluation of scientific knowledge requires a rise in explicity: time schedules and project formats become more fine-grained, cost and quality evaluations are introduced everywhere, and patent claims are applied for more widely and must be expressed more precisely.
- The distribution of knowledge-producing activities that includes many and different agents demands explicit rules of exchange, standards of cooperation and codes of sharing risks and profits between the agents of a network.
- The computerization of work and communication presupposes the existence and formulation of explicit rules and explicit models of the organization, its goals and its quality standards.

But in spite of this strong imperative of explication. I assume that the relevance of non-explicit knowledge will rise under the regime of distributed knowledge production. The limits of explication that already became visible under the regime of modern industrial production will be enforced. An unlimited process of making knowledge explicit will threat the balances and the scope of collective learning under the highly sensible regime of distributed knowledge production.

Non-Explicit Knowledge in Science and Technology

Science seems to be the empire of explicitness. But we have learnt by science studies that scientific knowledge is based on non-explicit knowledge. Ludwik Fleck (1935) demonstrates that scientific statements, assumptions about the effects of instruments, and interpretations of empirical observations are deeply embedded in the "thought style" of a thought collective. It is this group-bound and incorporated knowledge that Michael Polanyi further defined as "tacit knowledge," and with reference to that Thomas Kuhn coined the famous term "paradigm." This statement does not mean that science has always remained an art, as some post-modern thinkers assume, but that the advantages of modern science are necessarily interwoven with non-explicit knowledge. Polanyi expresses the paradoxical relation between implicit and explicit knowledge more sharply. He states that the process of formalization of all knowledge that does not exclude any element of implicit knowledge does destroy itself (Polanyi 1985: 27).

Modern sciences incorporate more and more technical instruments (Joerges and Shinn 2001). They change towards "experimental systems" (Rheinberger 1997) where implicit skills and explicit knowledge are closely interwoven with the "epistemological objects." Harry Collins (1974) had already demonstrated that the published explicit physical and technological knowledge is not sufficient, if one wants to replicate a successful experiment at another scientific laboratory; the replication must include a person who was a member of the scientific research group or who shared the practices of the group for a while as a visitor. The TEA Laser Set study emphasizes the necessity of shared collective experiences and of incorporated knowledge. Even in the most explicit science of mathematics, the proving and the formal examination of proofs with and without the help of computers is based on background knowledge of the mathematical culture that cannot be spoken out and be made completely explicit (Heintz 2000: 175).

If we accept the changes towards distributed knowledge production as described in the second part, then we must expect a rising consciousness of the tacit and implicit aspects of knowledge. Inter- and transdisciplinary cooperation cannot be performed by following algorithmic rules; but an enculturation process is required whereby the heterogeneous participants learn to know the tacit knowledge of the others, where they develop a shared language, and where they find a new community of practices. There is, however, also a strong force to make explicit more and more of the tacit assumptions in the process of scientification, but the timing between and the concertation of the different disciplinary packages of explicit knowledge needs a space where new tacit and collectively incorporated knowledge may grow.

Non-Explicit Knowledge in the Economy

Modern economy is said to be based on rational choices between goods whose values and costs can be made explicit. But especially in the field of innovation and technological choices it is evident that decisions are more likely to be following rules of thumb and organizational search routines than explicit rational calculations (Nelson and Winter 1982).

Under the regime of distributed knowledge production, the conditions of temporal planing and of cost calculation are worsening. A "circle of uncertainties" is coming up (Rammert 2002: 169) that limits the explicit calculation of risks and benefits. Enterprises try to cope with these uncertainties by building strategic alliances or by sharing knowledge and risks. As the success of an innovation is dependent on a growing bulk of environmental influences, for example, whether a policy of deregulation is introduced, whether norms of safety are increased, whether accidents happen or whether values and lifestyles change, it is only one strategy to intensify and expand the research function in order to make all of these uncertainties explicit and calculable. Another strategy of firms is to ask for all kinds of expert knowledge and to integrate it into the strategic planning. But both activities, the evaluation of the many, but different knowledge packages and the assessment of its weight in a dynamic model of future developments, can only be based on the tacit knowledge of "insiders" and on experiences of a mixed group of experts at the end. The method of a dynamic and collective scenario development tries to combine the use of explicit indicators and the implicit knowledge of a heterogeneous team of experts.

Non-Explicit Knowledge, the State, and Governance

Modern states are based on explicit constitutional rules and administrative laws. For a long time, they were conceived as the central agency of regulation. Specialized ministerial administrations defined the explicit legal frameworks of financial, economic, or science and technology policies. But policy studies could demonstrate that a formulated political program or a legal framework achieves the intended consequences, only if the conditions of its implementation, especially the interests and the experiences of the different collective actors in the field, are acknowledged (Mayntz 1993). New forms of governance were tested, like corporate governance, that divided governance between state authorities and private associations (Hollingsworth, Schmitter, and Streeck 1994).

When the actors who participate are growing, when the knowledge that is needed for formulating regulatory policies is radically distributed between the heterogeneous individual and collective actors, and when high technologies emerge with rapidity, then all kinds of knowledge that was until that time certain become quickly obsolete. The new knowledge that is until then highly needed receives the status of a rare, risky, and transitory good. Under these difficult conditions of distributed knowledge production, a kind of coordination mechanism is required that achieves both: It must maintain diversity of the actors and of their different knowledge perspectives, and at the same time, it must create a culture of trust and cooperation that allow s a kind of "distributed governance" that refers to explicit rules and to the implicit cultural model or "hidden curriculum" (Rammert 2002). We can observe a rise of procedures of mediation and the growth of interactive networks of innovation that integrate not only heterogeneous participants, but include even dissenting groups. These pluralist and democratic forms of network organization appear to become the new institutional answers to the problems of distributed governance.

Non-Explicit Knowledge and Complex Organizations

Bureaucracies as a prototype of formal organizations are defined by written rules and explicit procedures of rational operation, especially by the criterion of explicit membership (Weber 1968 [1921]). Empirical organization research, however, has demonstrated that informal rules, myths of rationality (DiMaggio and Powell 1983) and power that is based on knowledge about relevant zones of uncertainty (Crozier and Friedberg 1980) are more likely to determine the course of organizational development than explicit goals.

The complexity of organizations is actually raised under two aspects: internally, by the introduction of the new information and communication technologies (Sproull and Kiesler 1992; Orlikowski et al. 1996) that doubles, virtualizes and reorganizes tasks and interactions, and externally, by the increasing mass of actors and aspects that influence its performance from outside. New situations of a "distributed cooperation" (Rammert 1999b; Meister 2002) emerge and replace the Fordist work organization that is based on a functionally specialized division of work and the Tayloristic work organization that is mainly shaped by a hierarchical division of work planning and performance. Cooperation and interaction at high-tech workplaces are characterized by a high grade of agency of physical and software tools, a high grade of diversity of skills and competences, and a low grade of overall formalization. Governance is assured by pieces of explicit standards, routines, "boundary objects" and other borderline activities that mediate between the different knowledge cultures (Galison 1997; Star and Griesemer 1989).

The interactive networking between firms and the heterogeneous networking between organizations from different institutional backgrounds threaten to dissolve the explicit boundaries between organizations and institutional spheres. In the radically distributed world of virtual enterprises, for example, the clear-cut role and loyalty of an engineer is diffused by his cooperation in a development department that is commonly shared between two competing enterprises, and by his participation in a governance committee where norms of self-regulation are negotiated among industry, administration, and political representatives. It will be not a question of formal job descriptions, but of organizational culture and tacit dimensions of management whether this engineer shares the fresh attained knowledge with his original organization or uses it for influencing the governance committee, or takes personal advantage and sells it as a free consultant. The intermingling of expertise knowledge and of loyalties increases under conditions of distributed knowledge production and favors the rise of a flourishing consulting industry.

Finally, the considerations and findings of this section demonstrates neither that explicit knowledge will lose its relevance or even diminish under the regime of distributed knowledge production, nor that non-explicit knowledge is disappearing step by step. There is no zero-sum-game between explicit and non-explicit knowledge; both kinds of knowledge can grow and gain more relevance at the same time. There is one strong tendency in knowledge society to raise the level of explication in all fields when it is confronted with the rise of material complexity, when it has to cope with the growing discontinuity in the course of innovation, and when it has to integrate the increasing diversity of actors and their perspectives. Additionally, the employment of computer technologies and the progress of telecommunication strengthen the tendency to make knowledge more explicit. But the traditional limits of explication that were demonstrated at the fields of science, economy, state policy, and organization of work, do not disappear. There is another strong tendency to take care of the non-explicit knowledge. It gains rising relevance under the regime of distributed knowledge production. Non-explicit knowledge is a necessary condition

- for the creation and diffusion of scientific facts and technological artifacts in a growing multidisciplinary landscape.
- for pushing the technological innovation in times of high uncertainties,

- for the policy-regulation when the central authority of the state and when the certainty and disposability of knowledge are fading, and
- for the management of organizations that cross the borders towards virtual enterprises and network cooperation.

A Plea for a Policy of Knowledge Diversity and Collective Learning

In industrial society, the division of work produced an explosion of productivity. Big complex organizations and markets became the central institutional mechanisms of coordination. A regime of complementary and specialized differentiation of activities and institutional spheres was established to cope with the problems of raising the efficiency of and regulating the relationships among the production, distribution, and use of industrial goods. Policies of efficiency raising and economic growth were the adequate means on the road to a converging welfare society.

This road of convergence was shattered by new challenges that resulted from the oil crisis, the limits of growth, and the microelectronic, bio-genetic, and internet revolution. We are living now in a time of fermentation. Different modes of production and different regimes are developing side by side, sometimes in cooperation, sometimes in conflict with each other. There is no doubt about the new tendency, that the production, distribution, and use of knowledge has gained central importance for our society. In analogy to the division of work in industrial society, a parallel process—the division of knowledge—can be observed in actual society. We argued that this division of knowledge produces a growing mass of heterogeneous types of knowledge. It is produced at different places. But knowledge is a good of a different kind than cars or electronic circuits. It is intangible. Information can easily be distributed and copied, if they are once developed and packaged. Knowledge, however, grows only in actions and interactions. If knowledge production is dispersed over many places, if it is distributed between many different actors, and if it is split between heterogeneous perspectives, a policy of its accumulation, unification, standardization, and modularization seems to be rational according to the experiences with the industrial and economic processes. But it risks the destruction of those aspects that lend knowledge its particular worth: the richness of aspects and the relational character of knowledge will be lost.

The worth of knowledge rises, the more it is used and the more different are the aspects under which it is used, whereas the use of tangible goods lowers its worth by consumption. Therefore, a smart knowledge policy should encourage the diversity of actors and perspectives. It should cultivate the differences in and between the communities of practice. It should enable criss-crossing between different disciplines of knowledge. And it should keep open spaces and places where collective learning between heterogeneous actors can take place. A policy of quantitative knowledge growth should be complemented by a qualitative policy of knowledge diversity.

References

Beck, U., A. Giddens, and S. Lash (1994) *Reflexive Modernization*. Cambridge: Polity Press.

Böhme, Gemot, Wolfgang van den Daele, and P. Weingart (1976) "Finalization in science," *Social Science Information* 15: 307–30.

Böhme, Gemot, and Nico Stehr, eds. (1986) *The Knowledge Society*. Dordrecht: Reidel.

Button, G., ed. (1993) *Technology in Working Order. Studies of work, interaction, and technology*. London: Routledge.

Callon, M. (1993) "Variety and Irreversability in Networks of Technique Conception and Adoption," in *Technology and the Wealth of Nations*. C. Foray and C. Freeman (eds.). London: Pinter. 232–68.

Collins, H.M. (1974), "The TEA set: Tacit knowledge and scientific networks," *Science Studies* 4: 165–86.

Crozier, M., and D. Friedberg, eds. (1980) *Actors and Systems: The politics of collective action*. Chicago: University of Chicago Press.

Deleuze, Gilles, and Felix Guattari (1992) *Mille Plateaux. Capitalisme et Schizophrenie 2*. Paris.

Dewey, J., and A. Bentley (1949) *Knowing and the Known*. Boston: Beacon Press.

DiMaggio, P., and W. Powell (1983) "The iron cage revisited: Institutional isomorphism and collective rationality in organizational fields," *American Sociological Review* 48: 147–60.

Gibbons, M., C. Limoges, H. Nowotny, S. Schwartzman, P. Scott, and P. Trow (1994) *The New Production of Knowledge: The dynamics of science and research in contemporary societies*. London: Sage Publications.

Fleck, L. (1935) *Genesis and Development of a Scientific Fact*. Chicago: University of Chicago Press.

Foray, D., and C. Freeman, eds. (1993) *Technology and the Wealth of Nations*. London: OECD.

Freeman, C. (1992) "Networks of innovation: A synthesis of research issues," *Research Policy* 20: 499–514.

Galison, P. (1996) "Computer Simulations and the Trading Zone," in *The Disunity of Science: Boundaries, contexts, and power*. P. Galison, and D. Stump (eds.). Stanford, Cal.: Stanford University Press. 118–57.

—— (1997) *Image and Logic: A material culture of microphysics*. Chicago, Ill.: University of Chicago Press.

Gibbons, M., C. Limoges, H. Nowotny, S. Schwaxtzman, P. Scott, and P. Trow (1994) *The New Production of Knowledge. The dynamics of science and research in contemporary societies.* London: Sage Publications.

Grundmann, Reiner, and Nico Stehr (1997) "Klima und Gesellschaft, soziologische Klassiker und Außenseiter," *Saziale Welt* 47: 85–100.

Hage, J., and R. Hollingsworth (2000) "A strategy for analysis of idea innovation networks and institutions," *Organization Studies* 21: 971–1004.

Havek, F.A. (1945) "The use of knowledge in society," *American Economic Review* 35(4): 519–30.

Heintz, B. (2000) *Die Innenwelt der Mathematik: Zur Kultur und Praxis einer beweisenden Disziplin.* Vienna: Springer.

Helmstädter, E. (2000) "Wissensteilung," *Graue Reihe des Instituts Arbeit und Technik, Gelsenkirchen: 2000–2012.*

Hollingsworth, R., P. Schmitter, and W. Streeck (1994) *Governing Capitalist Economies.* New York: Oxford University Press.

Hutchins, E. (1996) *Cognition in the Wild.* Cambridge: MIT Press.

Joerges, B., and T. Shinn (2001) *Instrumentation: Between science, state, and industry.* Dordrecht: Kluwer.

Knorr Cetina, K. (1992) "Zur Unterkomplexität der Differenzierungstheoric: Empirische Anfragen an die Systemtheorie," *Zeitschrift für Soziologie* 21(6): 406–19.

———— (1999) *Epistemic Cultures.* Cambridge: Harvard University Press.

Latour, B. (1994) "On technical mediation: The messenger lectures on the evolution of civilization," *Common Knowledge* 3(2): 29–64.

Luhmann, N. (1977) "Differentiation of society," *Canadian Journal of Sociology* 2: 29–53.

Lundvall, B.-A. (1993) "User-Producer Relationships, National Systems of Innovation and Internationalization," in *Technology and the Wealth of Nations*, D. Foray, and C. Freeman (eds.). London: OECD. 277–300.

Machlup, F. (1962) *The Production and Distribution of Knowledge in the United States.* Princeton, N.J.: Princeton University Press.

Malsch, T. (2001) "Naming the unnamable: socionics or the sociological turn of/to distributed artificial intelligence," *Autonomous Agents and Multi-Agent Systems* 4: 155–86.

Malsch, T., ed. (1998) *Sozionik: Soziologische Ansichten zur künstlichen Sozialität.* Berlin: Sigma.

Marin, B., and R. Mayntz (1991), *Policy Networks.* Frankfurt am Main: Campus.

Mayntz, R. (1993) "Networks. Issues, and Games: Multiorganizational interactions in the restructuring of a national research system," in *Games in Hierarchies and Networks*, F. W. Scharpf (ed.). Frankfurt am Main: Campus. 189–209.

Meister, M. (2002) "Grenzzonenaktivitäten. Formen einer schwachen Handlungsbeteiligung der Artefakte [Trading Zone Activities]," in *Können Maschinen handeln?* Werner Rammert and I. Schulz-Schaeffer (eds.). Frankfurt am Main: 189–222.

Münch, R. (1990) "Differentiation, Rationalization, Interpenetration: The emergence of modern society," in *Differentiation Theory and Social Change: Comparative and historical perspectives*, J. Alexander and P. Colony (eds.). New York: Columbia University Press. 441–64.

Nelson, R., ed. (1993) *National Innovation Systems: A comparative analysis.* Oxford: Oxford University Press.

Nowotny, H., P. Scott, and M. Gibbons (2001) *Re-thinking Science: Knowledge production in an age of uncertainties.* Cambridge, Mass.: Polity Press.

Orlikowski, W., G. Walsham, M. Jones, and J. de Gross, eds. (1996) *Information Technology and Changes in Organizational Work.* London: Chapman and Hall.

Polany, M. (1985) *Implizites Wissen.* Frankfurt am Main: Suhrkamp.

Powell, W. (1990) "Neither market, nor hierarchy: Network forms of organization," *Research in Organization Behavior* 12: 295–336.

Powell, W., K. Koput, and L. Smith-Doerr (1996) "Interorganizational collaboration and the locus of innovation: Networks of learning in biotechnology," *Administrative Science Quarterly* 41(1): 116–45.

Rammert, W, M. Schlese, G. Wagner, J. Wehner, and Weingarten (1998) *Wissensmaschinen [Knowledge Machines].* Frankfurt am Main: Campus.

Rammert, Werner (1999a) "Inquiry into innovation—A Pragmatist's Conception of Technological Change." Unpublished paper. Madison, University of Wisconsin.

—— (1999b) "Routinen und Risiken verteilter Kooperation: Der OP als Beispiel für hochtechnisierte Arbeitssitutationen [Routines and Risks of Distributed Cooperation]," unpublished Research Project Proposal, Technical University Berlin.

—— (2000) Ritardando and Accelerando in Reflexive Innovation, or How Networks Synchronise the Tempi of Technological Innovation. Working Paper TUTS-WP-7–2000, Technical University Berlin.

—— (2002) "The Cultural Shaping of Technology and the Politics of Technodiversity," in *Social Shaping, Guiding Policy*, R. Williamson and K. Sorensen (eds.). Edinburgh: Edgar Elger Press.

Rheinberger, H.-J. (1997) *Toward a History of Epistemic Things. Synthesizing proteins in the test tube.* Stanford, Cal.: Stanford University Press.

Schimank, U. (1996) *Theorien gesellschaftlicher Differenzierung.* Opladen: Leske and Budrich.

Sproull, Lee and Sara Kiesler (1992) *New Ways of Working in the Networked Organization.* Cambridge, Mass.: The MIT Press.

Star, S.L., and J.R. Griesemer (1989) "Institutional ecology: 'Translations' and coherence Amateurs and professionals in Berkeley's Museum of Vertebrate Zoology, 1907–1939," *Social Studies of Science* 19: 387–420.

Star, S.L. (1996) "Working Together. Symbolic Interactionism, Activity Theory, and Information Systems," in *Cognition and Communication of Will*, Y. Engström and D. Biddhton (eds.). Cambridge: Cambridge University Press.

Stehr, Nico, and Hans von Storch (1999) *Klima, Wetter, Mensch.* Munich: Beck.

Stehr, Nico (2000) *Die Zerbrechlichkeit moderner Gesellschaften*. Weilerswist: Velbrück.

Tyrell, H. (1978) "Antragen an die Theorie der gesellschaftlichen Differenzierung," *Zeitschrift für Soziologie* 7: 175–93.

Weber, M. (1968 [1921]). *Economy and Society. Three volumes*. Totowa. N.Y.: Bedminster Press.

Weingart, P. (1983) "Verwissenschaftlichung der Gesellschaft—Politisierung der Wissenschaft," *Zeitschrift für Soziologie* 12(3): 225–41.

—— (1997) "From 'Finalization' to 'Mode 2': Old wine in new bottles," *Social Science Information* 36(4): 591–613.

5

Policing Society: Genetics, Robotics, Nanotechnology

William Leiss

The paper opens with the question raised by Grundmann and Stehr, as to whether "knowledge policy" may include "the aim of limiting, directing into certain paths, or forbidding the application and further development of knowledge." It then explores this theme with reference to contemporary developments in biotechnology and nanotechnology, where the objective of knowledge is to enable us to create and modify at will biological entities (including humans and combined species known as "chimeras"), as well as self-assembling mechanical entities, *ab initio* through recombinant DNA techniques. I argue that a new category of risks is created by the promised technological applications of these forms of knowledge, called "moral risks." which threatens the ethical basis of human civilization; these are also "catastrophic risks," in that their negative and evil aspects are virtually unlimited. The paper asks whether our institutional structures, including international conventions, are robust enough to be able to contain such risks within acceptable limits; or alternatively whether these risks themselves should be regarded as unacceptable, a position that would impel us to seek to forbid individuals and nations from acquiring and disseminating the knowledge upon which those technologies are based.

Introduction: "Eppur si muove!" ("And yet it moves!")

The background paper by Reiner Grundmann and Nico Stehr, "Policing Knowledge: A New Political Field" (2001: 1–2), poses "the question of social surveillance and regulation of knowledge." They suggest that "knowledge policy" may include "the aim of limiting, directing into certain paths, or forbidding the application and further development of knowledge." If scientific knowledge is included here, as I assume it is, this proposition will not be well received. One of the great founding faiths of modern society is that of the infinite benefits of the liberation of the natural sciences from the intellectual and institutional shackles of dogma, including religion; its inspirational image is that of Galileo before the Inquisition, forced to recant publicly his belief about earth's movement in space, but unyielding in his mind and certain subjectively of his ultimate vindication.[1] Anyone who seeks to challenge this faith is in for a rough ride.

Are there forms of knowledge about nature (including a technological capacity to manipulate nature based on them), now envisioned as practical possibilities in the foreseeable future, of which it may be said that they are too dangerous for humanity to possess? Too dangerous, at least, in the hands of that radically imperfect humanity in and around us, including its all-too-delicate veneer of civilization, which now seems prepared to seek that knowledge? And if so, is it even conceivable that one could argue for their suppression on the grounds that, once realized, they will inevitably be deployed, to ends so evil, running unhindered into the future, as to destroy the moral basis of civilization?[2] I at least am not ready to answer these questions— although they are being raised by some in the academic community, especially with reference to biotechnology. An editorial earlier this year in *New Scientist* (13 January 2001, see also pp. 5–6), commenting on the inadvertent laboratory creation of a virulent engineered virus that could be used as a weapon in biological warfare, said:

> There's also the problem that many biologists choose to ignore biotechnology's threats. . . . John Steinbruner of the University of Maryland, College Park, has suggested setting up bodies to oversee areas of biological research. Such bodies could question or even stop research, or decide if results should be published. As Steinbruner is well aware, his proposal strikes at the heart of scientific openness and freedom. But leaving things as they are is not an option. Biotechnology is beginning to show an evil grin. Unless we wipe that smile from its face, we'll live to regret it.

Here I wish only to ruminate on the conference themes with reference to a number of potentially catastrophic risks—risks having a dimension that calls into question the future of humanity itself—related to advances in contemporary scientific knowledge.

I define "catastrophic risk" as the possibility of harms to humans and other entities that call into question the future viability of existing animal species, including our own. Thus these are not only risks to the present generations of living animal species, but also to future (perhaps *all* future) generations of presently existing species. One well-known risk of this type is what has been called "nuclear winter," the threat of a pervasive environmental catastrophe that could follow a large-scale exchange of nuclear weapons between the United States and the former Soviet Union (now Russia), under the doctrine of "mutually assured destruction." The hypothesis of environmental catastrophe was based on the expectation that the earth's atmosphere would become loaded with particulate matter, blocking much of the solar radiation reaching the earth's surface, perhaps for a period of years (such an event is thought to have occurred following the impact of massive asteroids colliding with the earth).[3] In addition, of course, the huge doses of radiation emitted by these exploding weapons would have profound genetic consequences for plants and animals.

The Lords of Creation

Given the existing stockpiles of nuclear weapons, the risks associated with them still exist, although (in view of the political instability in Russia) it is difficult to know whether the probability now is greater or less than before. But new catastrophic risks are on the horizon, and these have a fundamentally different character that may require very different institutional responses from us. Their common characteristic, considered as basic and applied science and the technological applications made possible through them, is that they are all based on our latest understanding of biological systems through molecular biology. More specifically, their common scientific basis is the capacity to characterize complete genomes and to manipulate them by means of recombinant DNA techniques (or to create DNA-like mechanical structures).

The ultimate goal, already envisioned and set as an objective for research, is a knowledge of genomics so complete that living entities (and life-like mechanical entities) could be constructed, or alternatively deconstructed and then rebuilt and varied, *ab initio*. According to an article published in *Science* in 1999 (and cited in *The Scientist* 14(1): 12), researchers working with a microbial parasite sought to characterize and develop "an organism with a minimal genome, the smallest set of genes that confers survival and reproduction":

> But since each of the 300 genes found to be essential could have mul-
> tiple functions (pleiotropism), investigators had no way of finding the
> degree of redundancy and whittling the genome down further. The

next logical step: make a synthetic chromosome of just those genes to build a living cell from the ground up.

Considered in their human implications, I regard these developments as giving rise to a new type of catastrophic risk, which I have called "moral risks."[4] Gradations of being (inorganic and organic matter, plants, insects, animals, humans) are and always have been a foundation of humanity's ethical and religious systems. More particularly, "self-consciousness" has been regarded as the essential and distinguishing mark of a human being, uniquely; yet as illustrated in the following section we have, apparently even among some senior scientists, an inclination to experiment with "crossing" these dimensions of existence in an almost casual mood. In my opinion very great evils await us in going down that road.[5]

A Short List of "Catastrophic Risks"

1. There are risks from the use of future bio-engineered pathogens used as weapons or war or terrorism.[6] A recent review in *Nature* listed the following possibilities (Dennis 2001):

   a.   transferring genes for antibiotic resistance (e.g., to anthrax or plague, as Russian scientists have done) or pathogenicity (the toxin in botulinin, which could be transferred to *E. coli*), or simply mixing various traits of different pathogens, all of which is said to be "child's play" for molecular genetics today:

   b.   through "directed molecular evolution," especially what is called "DNA shuffling," producing "daughter genes" by shattering genes and then recombining gene fragments in ways that change the natural evolutionary pathways of bacteria;

   c.   creating "synthetic" pathogens, that is, "artificial" bacteria and viruses, by starting with a synthesized "minimal genome" which was capable of self-replication (a kind of empty shell), to which "desired" traits could be added at will; and

   d.   creating hybrids of related viral strains.

   These possibilities multiply as scientists begin publishing the complete DNA sequences of well-known pathogens: "[G]enomics efforts in laboratories around the world will deliver the complete sequence of more than 70 major bacterial, fungal, and parasitic pathogens of humans, animals, and plants in the next year or two" (Fraser and Dando 2001: 2). Scientists working in these areas point out that actually getting engineered viruses and bacteria to survive in the environment, and to be maximally useful as weapons of war and terrorism, would not be easy to do; moreover, defenses against them can be constructed. What we are faced with the advances in molecular genetics, therefore, is an increase in the risks (possible harms) of novel agents being used in these ways for nefarious purposes.

2. There are related risks from accidental or unintended consequences of genomics research, especially from the genetic engineering of viruses and bacteria, which could result from the escape into the environment of virulent new organisms, irrespective of whether these organisms were intended originally for "beneficent" or "malevolent" purposes.

There was a brief flurry of publicity earlier this year when Australian researchers announced that, in engineering the relatively harmless mousepox virus with a gene for the chemical interleukin-4, in an attempt to create a contraceptive vaccine for mice, they had accidentally made the virus exceptionally toxic: "The virus does not directly threaten humans. But splice the IL-4 gene into a human virus and you could create a potent weapon. Add the gene to a pig virus, say, and you could wreck a nation's food supply."[7]

3. There are risks to the "nature" of humans and other animals from intended or unintended consequences of genetic manipulations that either introduce reproducible changes into an existing genome (e.g., human or animal germ-line gene therapy), thus modifying existing species, or create entirely new variant species. For illustration here, I will confine myself to the example of "chimeras," that is, combined entities made up of parts of the genome of two or more different species, including of course humans. Some molecular biologists apparently already have done casual experiments inserting human DNA into the eggs of other animals and growing the cell mass for a week or so; and there is much speculation as to what would happen if human and chimpanzee DNA were crossed, since chimps share over 98 percent of human genes.[8]

4. The DNA of all species now on earth is composed of the same four chemical bases, abbreviated A, T, C, G, arranged into two pairs (A/T, C/G), that make up the "ladders" on the double helix of DNA; different combinations of the base-pairs specify one of 20 amino acids, which combine to form various proteins.[9] Some scientists are experimenting with adding more chemicals that would act as new bases, so that, for example, there would be six rather than four bases and perhaps three base-pairs. One of the scientists doing this work is Peter Schultz: "Schultz often says living things have only 20

amino acids because God rested on the seventh day. 'If He worked on Sunday,' he said, 'what would we look like?' "[10] The self-comparison between Dr. Schultz and God is interesting, to say the least.

5. Risks to organic life, stemming from certain possibilities inherent in the development of robotics and nanotechnology, were publicized in a now-infamous paper (April 2000) by Bill Joy, chief scientist at Sun Microsystems and creator of the "Java" script. Joy wrote:

> The 21st-century technologies—genetics, nanotechnology, and robotics (GNR)—are so powerful that they can spawn whole new classes of accidents and abuses. Most dangerously, for the first time, these accidents and abuses are widely within the reach of individuals or small groups. . . . I think it is no exaggeration to say that we are on the cusp of the further perfection of extreme evil, an evil whose possibility spreads well beyond that which weapons of mass destruction bequeathed to the nation-states, on to a surprising and terrible empowerment of extreme individuals.[11]

The link between nanotechnology and biotechnology is fascinating: Although the former works with intrinsically inert materials, it is seeking to turn them into a perfect analogue of a biological (self-assembling) system. One of the leading Canadian scientists in this field, Dragon Petrovic, has explained the quest as follows:

> In the future, he predicts, technicians will teach individual molecules and atoms to assemble themselves into wires and sheets of impeccable purity and thinness. . . . [Imagine] instruments made of compounds that are self- assembled, atom by perfect atom—materials so pure that they could never snap apart or break under normal conditions. . . . "Imagine [Petrovic says] the linkage to telecom—can we get DNA molecules to self-assemble into perfect sheets and wires only an atom thick, and then send electrons and photons to stimulate the DNA to do things—start growing; stop growing; assemble into certain geometric shapes? It's analogous to what a structure like bone does in nature, where the brain is the electronic device and the nervous system transmits the information."[12]

Bill Joy's essay already had explored the dark side possibly inherent in the quest for self-replicating nanotechnology machines; the internal quotation in the passage by Joy below is from a book by Eric Drexler, *Engines of Creation* (1986):[13]

> An immediate consequence of the Faustian bargain in obtaining the great power of nanotechnology is that we run a grave risk—the risk

that we might destroy the biosphere on which all life depends. As Drexler explains: "Tough omnivorous 'bacteria' [created by nano-technology] could out-compete real bacteria: They could spread like blowing pollen, replicate swiftly, and reduce the biosphere to dust in a matter of days.... Among the congnoscenti of nanotechnology, this threat has become known as the 'gray goo problem.' Though masses of uncontrolled replicators need not be gray or gooey, the term 'gray goo' emphasizes that replicators able to obliterate life might be less inspiring that a single species of crabgrass. They might be superior in an evolutionary sense, but this need not make them valuable."

Joy ends his essay with a plea for the urgent need to begin thinking about how to contain these risks. We will need, he thinks, a rigorous regime to oversee the technology's development and require that certain applications be relinquished; "enforcing relinquishment," he says, "will require a verification regime similar to that for biological weapons, but on an unprecedented scale."

One important point must be emphasized here, namely, that what has been just described are (hypothetical) catastrophic "downside risks," that is, the potential for very great harms to be done through some future technologies that are already on the drawing-boards. For each of these developments there are both "upside benefits," resulting from future applications of these technologies that could bring substantial benefits to us, as well as the potential for "protective" technological innovations that could mitigate, offset, reduce, or even eliminate at least some of the downside risks. To take the example of the engineering of viruses as bioweapons: As a counter to this threat (and also just to reduce the debilitating effects of viral infections on population health), research is under way in molecular genetics to develop new antiviral drugs that can block the infectious action of any viruses at the cellular level (preventing receptor binding, cell penetration, replication, production of viral proteins, and so on; Haseltine 2001; cf. Miller et al. 2001: 305–7). Considered as a totality, however, what these conjoined prospects do is to continually "raise the stakes" in our technological game with nature, whereby the new sets of risks and benefits reflect both, and simultaneously, the potential for an upside of hitherto unattainable benefits and a downside of hitherto unimaginable horrors. As discussed in a later section, this entire prospect increases the challenge to our social institutions to manage our technological prowess so as to realize the benefits and avoid the harms, and likewise increases the risk that we will be unable to do so.

What is Different Today?

There are undoubtedly other types of catastrophic risks, but those introduced above are sufficient for purposes of discussion! My main point is that these newer risks are fundamentally different in character from the case of nuclear winter, and the difference has to do with the distribution of knowledge and technological capacity relevant to them (thus requiring a very different institutional response). The technologies giving rise to the nuclear winter risk are controlled by just two nation-states and are maintained (for the most part, and until now) under a thick blanket of military security and secrecy, although the smuggling of nuclear materials out of the former Soviet Union is cause for worry. Both the essential theoretical knowledge, and the engineering capacity needed to turn that into weapons, is confined to a relatively small circle of experts and officials. Not so with the new technologies.

The catastrophic risk areas listed above stem from current research programs that are widely distributed around the world; moreover, the strongest drivers of them are private corporations, including the large pharmaceutical multi-nationals, acting with full encouragement, support, and incentives from national governments. Especially where the possible health benefits of genetic manipulations are concerned, the combined public-private interests are overwhelmingly support-ive, driving the research ahead at an accelerating pace Governments especially are enthralled with the economic significance of these new technologies, are competing with each other under innovation agendas to capture major shares of the corporate investments, and are loathe to stop and think about unintended consequences.

All of the characteristics of the knowledge and applications in these areas mean that it is extremely difficult even to think about controlling either the process or the results. For one thing, the knowledge is widely distributed among individual scientists; for another, it is widely dis-tributed among private actors (corporations) which have the option of moving their operations on a regular basis, seeking perhaps the least-regulatory-intensive national base on the globe (Might we expect H.G. Wells' *The Island of Doctor Moreau* to be replicated many times?[14]) Third, the technologies themselves become increasingly "simplified" and thus easier to hide, if necessary, the genetics technologies, for example, can be carried out in small laboratories almost anywhere. Sergei Popov, the Russian scientist who pioneered germ warfare research using recombinant DNA techniques, observed recently. "The whole technology becomes more and more available. It becomes easier and easier to create new biological entities, and they could be quite dangerous" (Miller et al. 2001: 304).

Fourth, oversight is inhibited by the lure of truly extraordinary economic and health benefits promised by the new knowledge and technologies. And fifth, just the astonishing pace of innovation itself today makes the prospect of control and regulation a challenge.

During the past year national governments have been scrambling to respond to just a few of the dimensions of these new risks. Most attention has been focused on human cloning, where a few rogue scientists have challenged authorities in various jurisdictions to "try to stop us," and laws prohibiting this technology are being passed rapidly. But this is a relatively crude technology, albeit one that excites public attention, and one wonders whether authorities will become complacent about their ability to control unacceptable technologies due to their experience with this case. (Meanwhile, there are increasing reports that many genetics scientists are "going underground," in the sense that they have stopped talking publicly about their research in progress for fear that public reactions will be hostile and will result in official steps to halt it.)

Among the scientists cited in this paper, two (Bill Joy and Ian Ramshaw) have called for urgent action under the Biological and Toxic Weapons Convention (BTWC; 1975), to provide explicitly for a global oversight effort over some of the new technologies and their applications described earlier. Unfortunately, and ironically in view of what was to happen only two months later, at a meeting of the parties in Australia in July 2001, the United States unexpectedly blocked the process of completing a protocol under the BTWC that would have made the Convention something other than a statement of good intentions, for in its present form it has no provisions for verification or compliance monitoring. The U.S. government has been pressured by its biotechnology industry sector not to agree to a verification protocol, under which inspections of laboratories and other facilities by international teams of experts would be carried out in all the signatory countries, because industry fears that its intellectual property and commercial secrets could be compromised. At the time of writing other signatories were considering whether they should proceed to complete the adoption of the verification protocol without U.S. support.[15]

Unfortunately, we know international negotiation to be at the best of times a tedious and protracted process, and there is reason to believe that in this domain it could be fractious and unsuccessful. This is because all of the technologies described represent frontiers of industrial innovation in which great multinational corporations and the national governments that protect their interests (especially the United States) have significant investments; both corporations and governments would be loathe to see those investments and the immense payoffs expected from them jeopardized by an international control regime. A recent article co-authored by a molecular geneticist and a specialist in the international convention on biological weapons has called for an urgent new effort to strengthen verification under the 1975 Convention and to enlist the biomedical research community in an effort to strengthen deterrence against the uses of bio-engineered organisms for war and terrorism (Fraser and Dando 2001: 4).

Conclusion

Now is the time for intensive exploration of the theme of policing science and to ask the following types of questions:

1. Can we characterize a set of new catastrophic risks, as defined here, related to the leading-edge technologies that are being developed?
2. Do these new risks have an essential character that will make them difficult to control, because the knowledge and the technologies will be so widely diffused?
3. Can these risks be confined to acceptable dimensions by the institutional means now at our disposal, including international conventions on prohibitions? If not, what new tools do we need, and how can we get them?
4. Do professional associations of scientists working in these fields have special responsibilities to assist societies in controlling these risks, and if so, are those responsibilities now being discharged adequately?[16]

What is at risk in this game, now, is the possibility that the tension between science and society will become both unmanageable for institutions and unbearable for individuals, in other words, that the destructive applications of our operational power finally will overwhelm the rest. This possibility arises out of the striking contrast between the pace of change in social and legal institutions (especially international agreements), on the one hand, and in new scientific and technological breakthroughs in the sciences, especially in genomics, including applications relevant to biowarfare and bioterrorism. In biowarfare genomics the pace is painfully slow and progress often remains ineffective even after decades of negotiation, as in the case of the Convention on Biological and Toxic Weapons. Bioterrorism genomics proceeds at a frenetic and steadily-accelerating pace.

To reduce the probability that change in the second will overwhelm our social and legal capacity to steer technological development away from the zone of catastrophic risks, it is necessary first to get agreement among influential social actors that this *is*, as described here, a momentous challenge that contemporary society cannot avoid. The first practical test of our resolve in this regard, I believe, is whether influential scientists can be mobilized in the cause, scientists who will reaffirm the need for new oversight structures, to be erected both within the practice of science itself and also in the relation between science and society. Hegel made a remark, I believe, somewhere in his writings, to the effect that only the hand that inflicts a wound can heal it. The wound here is the rupture with the dominant pre-modern relation of humanity and nature, governed by value-laden categories of being, and its replacement by modern science's purely operational orientation to the totality of the natural world.

I will not speculate here on what a healing of that rupture could mean now, at least, not in any "ontological" sense. But in a practical sense, as a matter of public policy, I think it is clear what is required—namely, that the practitioners of science join others in a program to try to bring our operational powers under the control and direction of social institutions that have universal validity, ones that correspond in sufficient measure with the common aspirations of humanity. It is my contention that today's dominant institutions do not have such validity and that, as a result, everyone on earth is at risk of having these powers become instruments in an Armageddon waged to the bitter end by contending social, ethnic, national, and religious interests.

What remains to be seen is whether the task as defined here can be widely recognized and grasped as such, while there is still time, and whether our scientific enterprise can be steered towards the shelter of a social compact having universal validity.[17] If it turns out that despite our best efforts this cannot be done, there will arise a set of other questions that, for now at least, are too abhorrent for many even to consider. These questions have to do with the possibility that, taking both "normal" human passions and human institutional failings into consideration, there may be forms of knowledge that, as a practical matter, are too dangerous for us to possess, and that our only choice is to renounce and suppress such knowledge or suffer the consequences. In mentioning them we go to the heart of the fateful compact between science and society that has set the course for the development of modern society from the seventeenth century onwards, under the program known as the domination of nature. It is likely that contemporary society is not ready to deal with them, at least, not yet.

Notes

1. Galileo Galilei (1564–1642): http://www.rit.edu/~flwstv/galileo.html
2. There is a practical argument to the effect that, since the development and deployment of such knowledge cannot be thwarted, the most prudent course of action is to superintend its progress closely, so that technological antidotes to the potentially most frightful and destructive applications will be ready before they are needed. I regard this as a strong and possibly definitive counter-position to the one posed here in the serious of rhetorical questions.
3. On asteroid risk (http://impact.are.nasa.gov/): "Statistically, the greatest danger is from an NEO [Near-Earth-Object] with about 1 million megatons energy (roughly 2 km in diameter). On average, one of these collides with the Earth once or twice per million years, producing a global catastrophe that would kill a substantial (but unknown) fraction of the Earth's human population. Reduced to personal terms, this means that you have about one chance in 20,000 of dying as a result of a collision."
4. "We encounter a state of moral risk when we pose certain options for ourselves, as goals which might be realized by using science to manipulate nature, that imply fundamental changes in the 'order of being' as it has been experienced by humans until now" (Leiss 2001: 267).
5. See generally ibid., pp. 259–68, where Mary Shelley's great novel, *Frankenstein* (1816), provides the basis for discussion.
6. The awareness on the part of U.S. officials that the bioengineering of pathogens using recombinant DNA techniques could pose new bioterrorism and biowarfare risks goes back to the beginning of the 1980s (Miller et al. 2001 80–84). Under the leadership of Sergei Popov, Russian scientists at "Biopreparal," the huge cover operation for the former Soviet Union's biological warfare research program, began carrying out this type of recombinant research at about this same time, creating among other things a "super-plague" germ by inserting the gene for diphtheria toxin into plague bacteria, as well as engineering viruses so that they would trigger catastrophic auto-immune responses in the victims Popov and his associates were not only interested in making lethal products, their experiments included attempts to manipulate moods though alterations in brain chemistry (ibid.: 300–304).
7. *New Scientist*, 13 January 2001. "Ian Ramshaw, a member of the Australian team, says [no one] could have foreseen that the altered virus would kill even vaccinated mice." The researchers were so alarmed by what they had inadvertently done that they first notified the Ministry of National Defense, then waited two years before publicly announcing and publishing their experiment, simultaneously calling for modifications to the international convention on biological warfare to include devices of this type. The original story is m *New Scientist*, 10 January 2001. See also Miller et al. (310–12).
8. Scott Foster, "Man-beast hybrid beyond talking stage," *The National Post* (Toronto, Canada), 22 August 2001: A16. "Last October, Greenpeace Germany dug up a patent claim for a human-animal hybrid, . . . U.S.-based

Biotransplant and Australia-based Stem Cell Sciences grew-a pig-human embryo to 32 cells before ending its life."

9. On DNA see the superb graphics and animation at http://vector.cshl.org/dnaftb
10. Andrew Pollack, "Not Life as we know it," *The National Post* (Toronto, Canada), 26 July 2001, p. Al 5 (reprinted from *The New York Times).*
11. Bill Joy, "Why the future doesn't need us," *Wired Magazine* http://www.wired.com/wired/archive/8.04joy_pr.html
12. Allen Abel. "The God of Small Things," *Saturday sight Magazine* (*The National Post,* Toronto, Canada), 21 and 28 July 2001: 34–7.
13. Online m its entirety at: http://www.foresight.org/EOC/
14. First published in 1896, this is the story of a rogue scientist who sets up a secret scientific research facility on a remote Pacific island in order to pursue vivisectionist experiments on animals and humans. The entire text is available at http://www.bartleby.com/1001/0.html
15. http://www.brad.ac.uk/acad/sbtwc; see especially G.S. Pearson, M.R. Dondo, and N.A. Sims, "The US rejection of the Composite Protocol: A huge mistake based on illogical assessments," and G.S. Pearson, "Why Biological Weapons present the Greatest Danger," at http://www.brad.ac.uk/acad/sbtwc/evaluation/evalu22.pdf
16. The 1975 Asilomar Conference that established some early ground-rules for DNA research at the initiative of the scientific community itself, had a 25th-anniversary meeting in 2000. At least according to one report, some senior scientists today are doubtful that the "Asilomar model" will prove to be useful in the future for the oversight of problematic applications of DNA research, particularly because of the enormous pressure of commercial interests that has developed in the meantime. See *The Scientist* 14(7): 15, 3 April 2000.
17. There is not time here to develop this concept adequately. Here it must suffice to say that "universal validity" is not an absolute, in the sense that every person must "buy in," but rather is some common orientation that can attract and hold the support of the most influential and enduring cultural traditions around the world.

References

Dennis, Carina (2001) "The bugs of war," *Nature* 114 (17 May): 232–5.

Drexler, Eric (1986) *Engines of Creation: The coming era of nanotechnology.* New York: Anchor Books.

Fraser, C.M., and M.R. Dando (2001) "Genomics and future biological weapons: The need for preventive action by the biomedical community," *Nature Genetics,* advance online publication, 22 October (http://nature.com.anthrax).

Grundmann, Reiner, and Nico Stehr (September 2001) "Policing Knowledge: A new political field." Kulturwissenschaftliches Institut, Essen, Germany.

Haseltine, William A. (2001) "Genetic traps for viruses," *Scientific American* (November): 56–63.

Leiss, William (2001) *In the Chamber of Risks: Understanding risk controversies.* Montreal, Que.: McGill-Queen's University Press.

Miller, Judith, Stephen Engelberg, and William Broad (2001) *Germs: Biological weapons and America's secret war.* New York: Simon and Schuster.

New Scientist.

6

The New Human Zoo

Jeffrey Klein

What are the rules in the human zoo? Suddenly we're anxious because we sense that the current rules aren't strong enough to prevent the animals within us from breaking out. Science seems on the verge of unleashing our deepest desires. Paradoxically, the power to redesign our genes has some of us worried about our survival. Are we the last natural people?

Such fearful questions are, I believe, well-founded. Faustian appetites are indeed increasing among producers and consumers in the rapidly developing knowledge market. Traditional mediating forces seem inadequate, if not complied. Our species is poised—via technology, science, industry, government and the media—to transform itself. The creatures in the human zoo are likely to change faster than the posted zoo rules.

Some people (particularly academics) don't like the rules that are *already* posted, so they imagine a more regulated world, one more controlled by "good" values. But the market for optimizing human nature will fundamentally be run by capitalist and libertarian rules. Science will take a back seat to commerce, and social science may be left far behind in its own foggy idealism.

Thus a key issue for those interested in the governance of knowledge is coming to terms with powerful, harsh, market-based truths. That is the main point of my paper, though I'm also going to get at darker, older forces hidden in the new equations.

To illustrate my points, I'll start with bioinformatics, head out to weapons in space, then circle back to bioinformatics because (1) it's in today's news; (2) it's likely to remain in tomorrow's news because that's where both the big public and the big private money is heading; and (3) the science and technology of altering nature bring the harsh Hobbesian logic of human nature into high relief. I believe that something bigger than money is motivating us, but I'll come to that, paradoxically, when I talk about the most entrepreneurial animals in the human zoo.

So, first, why will bioinformatics attract so much money? Because of the size and importance of the market. With the advent of bioterrorism, the U.S. government now has a reason to sponsor advanced research in this field. Rich and poor alike will cry out to have their bodies protected from terrorist toxins. The capital markets were already hyper-focused on biotech because they know that first the rich, and then everyone else, will pay whatever it costs to have their germ lines rid of disease.

How might such developments proceed? Once the bugs are worked out in mice and then primates, human trials will take place among the desperate, terminally ill. Both government and industry researchers will actively experiment with the biological bases for various pathologies.

Sooner rather than later, the elites will start paying whatever it costs to improve their health and increase their longevity. Drugs and implants will first be customized for those who can afford them. Sooner or later, these same elites will selectively breed for taller tots with robust immune systems. Bigger breasts and better skin may also be part of the pre-implementation embryonic packages (call them PEPs) offered in libertarian countries that want to grow rich and strong themselves.

Does anyone seriously think that science will check its forward, breathless momentum? Science is now driven by technology, which is driven by industry, which will entice the mass market to demand what the rich are getting once the premium market starts saturating. Government will certainly seek to maintain a central role as regulator, but who doubts that at the end of the day the governing parties will work with their financial backers to guide an orderly, safe and profitable release of new potentialities within the human species?

Most critics underestimate the moral force that supporters of opt-in eugenics will be able to muster. In the human zoo, creatures fundamentally seek freedom not just in self-improvement, but also in moral decision-making. Those who want to upgrade their breeding stock will say: Stay in the old world if *you* want, but I prefer that my children have a better chance of living longer, happier, fear-free lives. The majority will declare a new form of independence. Everyone's offspring has the natural, inalienable right to be born superior.

Before I flesh out my predictions and outline possible political grounds for constructive governance, I'd like to give you a few data points about my background. For twenty-five years. I've been an investigative reporter and editor specializing in political exposés. Two years ago I made a complete career change. (Americans seem to do this sort of thing frequently.) I started a software company that specializes in high-dimensional data analysis and predictive modeling. When I taught courses at Stanford and UC Berkeley, they drew upon my experiences in journalism, but equally upon my formative intellectual training at Columbia University in the late 1960s.

I've always been prone to predictions. It's unscientific not to predict outcomes, but my own drive predates this rationalization. Over the years I've realized my limitations. The future isn't just hard to know, it's fundamentally unknowable. We grasp a few data points and extrapolate in a straight line, anchored by our original observations. These are two of the human animal's most common and costly cognitive errors—anchoring and extrapolation.

Let me tell you about my own intellectual anchoring so that you can put me in my place. I came to college with an ability to memorize facts and calculate, period. Having escaped a backwoods zoo, I wanted to study human nature in the modern metropolis—and therefore I majored in psychology. But Columbia's department at that time was "experimental"—which meant that it was empirical and reductionist. Rats in cages don't tell you much about human behavior unless, perhaps, you can pinpoint and analyze changes in their brains at the moment they are learning. That's what the most experimental behavior science departments are doing today.

In my era, the best scholar of human nature in the university was Lionel Trilling and I was lucky enough to spend my senior year studying with him in a small seminar as he prepared his final book, *Sincerity and Authenticity.*

Trilling was officially a professor of literature, but he dissected a wide range of cultural texts (including key narratives in philosophy, psychology and sociology) as a way of comprehending "the moral life in process of revising itself." His final work described how the modern European personality emerged in the seventeenth century and then immediately tried to demonstrate its "sincerity": saying honestly what you think and feel so that you will be admired and trusted. This ethic of sincerity was gradually usurped by the twentieth-century ethos of authenticity: acknowledging a baser, more violent self struggling to dominate a universe that yields no firm truth or guidance.

A traditional Anglophile with little patience for social theory, Trilling personally disliked the cultural artifacts of authenticity, but still he recognized that the modern self was seeking liberation from the etiquette of sincerity, which it felt to be a constricting social norm. From Trilling I learned both to admire and to distrust the human mind. The subtlest of intellectuals, he gazed with suspicion upon his own breed. He showed us how even the revolutionary masterpieces of each era conformed to deep conventions. In the late sixties, the most progressive thinkers (e.g., Marcuse and R.D. Laing) were urging the young to escape from their social selves—and Trilling felt that this led to adolescent madness. He deemed it dangerous advice —and vulgar thinking.

Trilling's inquisitiveness influenced me more than his instinctive caution. Not so long after graduation, I helped launch *Mother Jones*, an investigative magazine that has repeatedly exposed political corruption and corporate malfeasance. For example, we were the first to expose corporate dumping—how multinational corporations sold overseas hazardous products that had been banned m the United States. In my second *Mother Jones* stint a decade later, as editor-in-chief. I specialized in character assassinations of powerful politicians—for example, the speaker of the U.S. House of Representatives, Newt Gingrich, and presidential candidate, Senator Bob Dole.

Disclosures of illegal behavior or hypocrisy or (best of all) both can help humiliate and evict bad actors. You can help drive politicians out of office and corporations into court. In a country as anti-statist and envious as the U.S., the public is willing to think ill of its leaders and clamor for their dismissal if a sufficient, fact-based pattern can be developed. A fear of watchdogs can, in turn, help govern politicians and businesspeople.

Still, in both of my stints at *Mother Jones*—separated by a decade at Knight Ridder, one of the United States' larger newspaper corporations—I retained Trilling's suspicion of reformers. Critics of conventional society weren't, to my mind, automatically good. Often they selectively chose facts to support their predispositions—a common enough vice. Much worse to my mind was their betrayal of a core progressive principle: empiricism. They averted their eyes and minds from the failures of the social experiments they advocated, so they could continue to feel good about themselves and their moral intentions.

As you can tell, my own habit is to challenge those who assume authority. Those who would police knowledge frighten me. When someone speaks about the need for tighter governance, my initial instinct is to suspect his or her motives. Like many Americans, I seek freedom from any form of interference. If anyone seeks authority over me, I want the power to approve or revoke it.

Let me pause for a moment to define the roles of the authority-seeking players I've chosen to discuss. To simplify: The politician's role is to mediate between the powerful and the populace. The mainstream media's role is to gather audiences for marketers. The media will certainly sacrifice one marketer or another if it helps them draw a larger audience and enhances their apparent power to influence events. Both sets of mediating actors (politicians and the media) try to make themselves seem more independent and more important than they are. In fact, economic elites are now at the top of the First-World food chain by virtue of their capital *and* the widespread social belief that they are winners.

Like some social scientists, many journalists in the United States belong to an anti-elitist elite. Print journalists (my former cadre) aren't the best paid, but they tend to initiate the stories that the other media amplify with video and sound and emotion. Among print journalists, high status is accorded to those who expose and depose other elites. I've had the experience both of attacking and sponsoring technological elites.

For example, when I worked at the *San Jose Mercury News,* the newspaper of Silicon Valley, I researched and wrote a serialized, techno-thriller whose underlying purpose was to expose what I thought was Silicon Valley's darkest secret: the American government's black budget to dominate space with offensive weapons. Fifteen years later, I still believe that the United States intends to pacify space in the same way that it tamed the American West: Play by our rules, or else.

When the story was subsequently published as a paperback, it received more than its fair share of publicity. But at the book party held in Silicon Valley, and promoted heavily in its newspaper, virtually the only people who showed up were the psychiatrists and psychologists who were privately counseling the weapons makers. The secrecy architected by the government—no one part of the project knew what the other parts were doing—apparently troubled a few of the participating scientists and technologists. Still, they were eager to experiment on the horizons of knowledge in their particular specialty—e.g., nuclear-powered lasers in space. The rules of compartmentalization (*sehr Deutsch, ja?*) freed them from overall moral responsibility.

I pushed deeper, into the nearby nuclear weapons lab, Lawrence Livermore, financed by the U.S. government and administered by the University of California. As with my initial reporting, only the very top experts who had retired would talk. In their youth, the brightest and most creative had broken old physics' paradigms to make new weapons, once out of the loop, a few were willing to break the rules of the culture they no longer depended upon.

Among the younger physicists, however, the drive was not unlike the culture of the mafia where new recruits "make their bones" by killing someone. At the weapons labs, you "make your bones" by having ideas compelling enough to warrant an actual nuclear test—in that era, a test underground in the name of refining the safety of the tools that defend democracy.

At Lawrence Livermore Lab, there are many levels of security denoted visibly by the color-coded badges everyone wears all the time, but the highest status was accorded those whose basic science breakthroughs had led to the actual, experimental detonation of a big one—especially one that promised greater, more controlled destruction.

This is why I believe Professor John Polanyi's argument, which graces this volume, doesn't deal with the fundamental problem of how to govern science. He thinks that scientists shouldn't be forced to justify their research grants by proposing useful applications. Like a true Nobelman, Polanyi believes we'd have a better world if the brightest people were funded to experiment freely. But in my experience, governments will happily fund "basic science" in strategic fields where they want to maintain a lead. Both parties, in fact, share the drive for useful applications—it's a power drive. For scientists, power is best achieved through status, which is conferred by the testing of predictions derived from their new, grand theories; for nations, power is achieved through demonstrations of applied force.

Years after my Star Wars' reporting, when I was back at *Mother Jones*, I found a similar ethos and similar power drives within the budding biotech labs. All of the research is done in the name of remediating illness or eradicating disease—the biological equivalents of the U.S. Defense Department. But most of the Eros goes towards creating potent new life forms.

When I visited the Geron Corporation in Menlo Park (in the news now because it holds a lead on stem cell research) I peered at the first apparently immortal cells. Geron had cloned a gene that produces telomerase, an enzyme that keeps rebuilding the ends of chromosomes. Then the scientists transferred this gene into a young, healthy cell—and, voilà, the cell hadn't stopped dividing, nor had it mutated into a cancerous form.

This breakthrough promised a host of potential therapeutic applications, but Geron at that time was focused on three. The first: a possible cure for cancer. The flip side of discovering how cells perpetually divide could be discovering how to make malignant ones stop. If a cancer cure proved elusive, or too distant for impatient capitalists (who'd already invested $118 million), Geron intended to use the discovery to treat diseases associated with aging—e.g., retinal decay, hardening of the arteries, osteoporosis. And if that failed, a fallback revenue plan was to use this technology to treat wrinkles.

Geron patented the use of telomerase before it knew what the biotechnology might actually be able to do, a pattern likely to be repeated frequently in the coming decades. Prospective patents are crucial to fledgling companies frantically racing to become leading players in the new "eugenics revolution."

Whether you call eugenics "eugenics" or "human genetics," the impulse still remains to upgrade the stock of the human race. The big change is that the global marketplace and post-modern individualism are much better equipped to actualize this impulse. Why? Because the impulse to upgrade the stock of the human race is driven by *individuals,* who want to upgrade themselves, or their children. It isn't that we all want to improve the human race, but rather than many people want to climb as quickly as they can, by whatever means possible, to the perceived top of the human race. This will spur competition (and conflict), but as everyone tries to climb, the top itself will get higher.

Robert Wright, perhaps the best contemporary writer on evolutionary psychology, has articulated a key difference between the new eugenics and the old eugenics championed by a wide range of social engineers from Oliver Wendell Holmes to Hitler. The old eugenics aspired to upgrade the whole race and advocated doing so by centralized control. The new eugenics is being driven by individual aspiration, and will happen in a decentralized way, unless centrally prohibited.

The changes wrought throughout the human race (changes so vast they lie beyond our current reasoning capacity and imagination) will come about as mere byproducts of each individual's unleashed desires. These individual desires will be widely supported by many voters in many countries as long as genetic upgrades are marketed for a long time as a way to curtail or cure disease, and remain optional.

Still, the human animal's eagerness to seize the future is shadowed by a widespread fear that the future might seize us. Will our intellect unleash monstrous egoism? Will we become so brutally efficient that we lose our humanity? Some evolutionary psychologists, such as Robert Wright, aren't so fearful. Asserting that biological evolution and cultural evolution are essentially the same machinery driven by the same fuel in the same direction, they expect increasing complexification—intelligent brains making imaginative choices to ward off entropy. For example, some individual aspirants will recognize how maladaptive the fiercer side of human nature can be in the modern social environment, and they will choose to become mellower. They'll want their children to be endowed with the Buddha gene.

What's the early news about whether the benefits brought on by bioinformatics will exceed the dangers? Right now our businesslike attitude toward the transformation of human nature is punctuated with bursts of hysteria. In the U.S., few people paid much attention to biotechnology until Dolly. Then, three days after the cloned sheep was introduced to the public, President Clinton immediately urged a panel to consider the expansion of federal powers over human embryo creation. Biotechnology exploded into our consciousness in a Frankensteinian fashion, a news cycle repeated in the U.S. in the debate over stem-cell research and then bioterrorism.

President Bush's recent decisions show which way the political winds are blowing. On stem-cell research, Bush essentially turned his back on right-to-lifers, particularly Roman Catholics, and sided with moderates. His secretary of health and human services, Tommy Thompson, is a Catholic who had a strong pro-life record when he was governor of Wisconsin. But Thompson reportedly lobbied for this decision to permit some federal research funding, not least of all because the University of Wisconsin was the first to grow and sustain embryonic stem cell colonies; it holds the key patent.

Even Bush's compromise position is unlikely to hold for long because if other, less religious countries gain a practical advantage, American industry and the American public will wage a holy war over choice. Rules and regulations will certainly arise (particularly about safety and privacy), but ultimately those who want to improve themselves will win. The U.S. government has a long tradition of developing markets for aspiring individuals.

On bioterrorism, the U.S. government now, appropriately, has a green light to combat current threats and prepare for future "worst case" scenarios. That means developing biodefense systems for genetically engineered strains of disease. The mind boggles. Defecting

Russian scientists have described elaborate secret labs in which they created decoy diseases whose likely treatment with antibiotics would release even more destructive viruses. If you find it to hard imagine such strategic pathology, just think of a bomb exploding, then as rescuers arrive, the real weapon is detonated. Defending against maniacal bioweapons means that American researchers in federal projects will have to understand all varieties of mutations, which means that they will help advance this dark science as well. The U.S. government will also encourage the private sector of the bioinformatics industry by, for example, relaxing certain regulations and increasing tax incentives.

What's driving the biotech revolution? Fundamentally we are all optimizing creatures. The human instinct for status competition seems universal, though it expresses itself distinctly in different cultures and eras. Status drives come into conspicuous play when a significant percentage of the population has the resources to play the game. Sometimes the available resources and the cultural forms combine to accelerate the ability to aspire. Thai's true for twenty-first century America.

For many Americans, progress itself has become a god again, an attempt to reconcile prior religious beliefs with the new faith in technology to improve everything, including our economy. The more successful the West is with its project of perpetual self-enhancement, the more likely it is that disaffected, envious countries will try to destroy our technological and economic superiority in the name of their pure old god. Imperial optimism attracts messengers bearing death.

Ironically, bioterrorism will intensify the use of science and technology to enhance ourselves. But this project will play out in a pacified world as well. How? Here's one guess: Genetic science will come up with tests that tell which embryos are prone to develop cancer prematurely. Diagnosing an embryo should prove easier and cheaper than reversing disease in a fullygrown adult. Given a choice between fetuses that have perhaps twenty years' difference in their likely life spans, the business and cultural elites will be the first in line to select.

Those upwardly mobile parents who select little Johnny Over little Janie because doctors say he's destined to be remarkably robust aren't, at first, likely to tell their neighbors. Legal and moral reprobation aside, they'd probably want Johnny's vivacity to be seen as a natural reflection of their own virtues. Also, these early adopters won't want to be blamed if unintended genetic consequences occur, which they will. But sooner rather than later, improvements will be verified, word will leak out, and everyone will want to keep up with the new, improved Johnny Joneses.

Will these new, improved people feel divorced from us in some indefinable way? Perhaps their longing for old-fashioned connectedness will be tinged with narcissistic contempt. Because they will be physically superior and free from ordinary fears, they'll feel superior as well. They will resent the social and economic burdens imposed upon them by those not so well-engineered.

Social and political scientists will have their hands full—one hand trying to contend with the practical governance problems created by the new gene gap; the other engaged in hand-to-hand combat with more entrepreneurial academic departments. You all know the old joke: Why are academic fights so vicious? Because the stakes are so small. That situation has changed today for the computer and biological sciences, and will change much more drastically tomorrow for bioinformatics.

Why "bioinformatics" instead of "genomics"? Genomics is already being eclipsed by proteomics. It seems that proteins, not genes, may be the key to our reconstruction. And protein structures are much more complicated than genetic ones. So scientific elites seeking maximum power will require facility with software that helps them automate the extraction of useful information and the building of better predictive models that take into account all sorts of environmental interactions. A mastery of bioinformatics will thus be more crucial than an education in mere molecular biology.

Bioethical rhetoric notwithstanding (and prepare yourself for high-minded reports from professional ethicists), the trustees and the administrators of the top universities will side with the hard sciences. Why? They won't want to limit their royalties. If the pro-life U.S. secretary of health, Tommy Thompson, won't put up serious opposition, who will? The bioinformatics professors and graduate students will certainly want the means to carry their new ideas to fruition—and they'll migrate away from any institution that constricts them. At the private and the government-sponsored labs, there will be an irresistible drive to detonate the biological big one—which would be attacking death itself.

Most modern people believe that humans die from disease. In fact, diseases may be present in the newly cold bodies of the dead, but the real cause of most death is cellular aging. Our cells are born with instructions to divide only a certain number of times. The maximum number of divisions—the upper limit of life, about 125 years for humans—is called the Hayflick Limit, after its discoverer, anatomist Leonard Hayflick.

When I interviewed Professor Hayflick on public radio, I was surprised to hear his belief that the Hayflick limit will never be broken. In other words, we may keep increasing our life expectancy (e.g., more people living to be 100), but not our life spans (a maximum of 125 years). He may be right, but that won t prevent the best and the brightest scientists, capitalists and technologists from trying to rewrite the biological instructions with which we're currently born—especially now that we have isolated the body's own fountain of youth: embryonic stem cells.

As I mentioned at the outset, two years ago I started a software company, and so have wrestled with the powers that be. My venture into the heart of capitalism and high tech has confirmed many of my prior prejudices and destroyed others. I've learned that technologists are perhaps the most creative and destructive people I've ever met. They love building things. And they love destroying the inferior things that someone has built before them. Joseph Schumpeter, of course, saw this "perennial gale of creative destruction" as an essential dynamic of capitalism.

How to govern the new elites? At their core, most are rational materialists. They are profoundly utilitarian. Many of my best engineers are recent immigrants who still uphold their religious traditions, but not because these belief structures retain a claim on their brains. The main continuity between their past and our future is a shared desire to be useful. In the American melting pot, everyone seeks to be useful. Sometimes this urge is hard to discern beneath our competitive individualism. But the ideal to be of service to a larger community—the desire to realize our fundamentally social nature—is hardly dead.

In the United States, at least, this desire is not likely to be fundamentally stimulated, shaped or curtailed by the state. Anti-statism runs deep in America—it's a main reason that socialism has never gained any political traction. Our technologists sometimes look to the government for money, rarely for cultural or moral guidance. Entrepreneurs who deride entitlement programs feel the government should sponsor new markets—then get out of the way.

If governance of the new elites seems difficult, governance of the super-elites will require quantum leaps of willpower. The top mega-entrepreneurs, like Jim Clark, work and play with an open contempt for social rules. They often feel isolated from those whom they manipulate. Everything seems possible, nothing real. The mega-entrepreneur has already fulfilled every conceivable physical need, yet still feels unsatisfied. So he seeks a single solution for his insatiability: maybe he'll feel good if he can impose his drives upon the world.

When I wrote about Jim Clark for my old newspaper. I likened him to Captain Ahab in Melville's *Moby Dick*. Like Ahab, he has contempt for his financiers, who he deems mere merchants. Clark is after a much bigger fish: vengeance against nature for human vulnerability. In Ahab's case, it was the loss of his leg; Clark, like all of us, has his own vendettas, which are well documented in Michael Lewis' book, *The New New Thing*. As Hobbes understood, loss registers with us much more profoundly than gain. The most fearless biotech entrepreneurs, like the most fearless scientists, will want to dominate death itself.

Could the universities act as a brake on such Faustian ambitions? Unlikely. World-class research universities like Stanford try to stoke the ambitions of the Jim Clarks in order to benefit from their largesse. Clark made a $150-million gift to Stanford to fund a new bioinformatics department, and then he publicly suspended $60 million of this pledge in a *New York Times* op-ed article protesting President Bush's go-very-slow approach to stem cell research. Initially Clark was going to throw a $90-million tantrum, but Stanford's president (like Clark, a former engineering professor at the school) convinced the philanthropist to withhold a lesser sum in his protest against anti-science politics, but still give enough to erect the building that will bear his name.

Even if successful entrepreneurs weren't calling the shots, would-be entrepreneurs on the faculty will protest any strictures on them. If the academy attempts to govern the distribution of knowledge, technologists will be the first to subvert these rules. The Internet refrain that "information wants to be free" could be revised to say, "information seeks transmission." We are communicative creatures.

Another suggestion floated by a contributor to this volume—the idea of hiding scientific controversies—frankly appalls me. Democracy should be nourished at all levels of learning. A better way to govern exploding knowledge is to let as many people as possible participate in its creation and application. If mass individuation is a dominant social trend of our era—everyone desiring to be different than everyone else—then dialogue among different aspirants may temper the perpetual war of all against all that Hobbes postulated without resorting to the authoritarian state he favored.

The promises of egalitarianism haven't been realized, but it's the best social ethic we've hatched to date. Because egalitarianism promises fulfillment to different types of believers (including political enemies), it can mitigate the bioinformatic war of all against all. Egalitarianism says: Let's elect politicians who will craft rules that allow each of us to be the agents of our own destiny.

My greatest fear is not that new knowledge will be used to create permanent genetic castes, but rather that in the headlong rush to transform our species, we will destroy what is precious about ourselves. Pioneers have a nasty history of destroying first and thinking second. This pattern apparently goes back at least as far as the Clovis hunters who swept into North America about twelve thousand years ago and rapidly hunted all the mammoth mammals to extinction.

What could we destroy today? The point is: We don't know. We have no idea. But the enormous progress we've made in the past few centuries has been accompanied by a vulgar egoism eager to exploit everything as quickly as possible.

There's a joke you've probably heard in one cultural form or another. The top English engineer and the top French engineer meet again and again and again to plan the channel tunnel that will unite them. Finally, after years of planning and testing, the French engineer tells his English counterpart: "You've completely convinced me that this will work in practice, but I'm still not convinced that it will work in theory."

I don't know how the politics of knowledge will play out in theory. My guess, as I stated at the outset, is that it will play out in practice faster. We will change our nature, and then see how these changes affect our social behavior. Some of the German scholars who sponsored this volume fear a chaotic future because they are planners; chaotic change could curtail the order and the freedom that they cherish. Clear, rational, bureaucratic governance feels more comfortable to them because it flows naturally from their prior obedience to the church and the state, and honors the familiar virtues of duty and obligation. But many of my American co-workers want to bring on the future as fast as they can because they are competitive optimists who believe that bad ideas will fall by the wayside and hard chargers like themselves will emerge in charge. A good knowledge society, in their minds, basically satisfies as many human desires and aspirations as possible.

The shifting of humanity's tectonic plates—be they proteomic or socioeconomic—will shake up traditional belief structures. Strange affinity groups may form on opposite sides of newly discovered fault lines. For example, fundamentalists on the right and environmentalists on the left may find themselves agreeing about the urgent need to stop further experimentation and preserve life as we know it. And they may be opposed by a heterogeneous mass championing the drive towards super-homogeneity under the banner of seizing our destiny.

Both the new aspirants and the old tribalists will find themselves facing a core human dilemma: an inability to comprehend the decisive moment in which we perpetually find ourselves. Our desires press us forward, our fears draw us back. The honor we all seek is barely-clothed ambition.

Faced with a man-made state of supernature, many citizens of the planet will support leaders and institutions that promise to protect each individual's freedom. Freedom of moral choice will be cherished as fiercely as freedom of choice within economic markets, and many people will conflate the two. Crudely put: Capitalism has unleashed deep human drives. Technology in general and biotech in particular promise to give consumers what they desire. Governance structures in the knowledge market will remain weak because neither producers nor consumers want to slow down, and competitive pressures won't let them, anyway.

Part III

Case Studies on the Governance of Knowledge: The Ineffective Governance of Science

Introduction

J. Rogers Hollingsworth

Nico Stehr is to be commended for placing the subject of the governance of science on our research agenda. For some years, various social scientists have been studying the governance of different sectors of capitalist societies (Campbell, Hollingsworth, and Lindberg 1991; Hollingsworth, Schmitter, and Streeck 1994; Hollingsworth and Boyer 1997). However, these perspectives have not been applied to the governance of knowledge and science. The problem of conceptualizing the governance of knowledge and science is complex, but hopefully this section can help to clarify some of the issues which we should confront.

As some of the papers in this volume point out, we do not yet have a good understanding of what we mean by knowledge. There is traditional as well as modern knowledge. Michael Polanyi (1958) differentiated between tacit and codified knowledge. In addition, there is the problem of defining science. Some draw the distinction between applied and basic science (Hage and Hollingsworth 2000). Are knowledge and science one and the same? Some contend that biological scientists rarely discuss knowledge; they simply focus on doing experiments and add information about various phenomena (interviews with Gerald Edelman and Ralph Greenspan 1998). In short there are many different definitions of science and knowledge. Due to limitations of space there is no effort in this chapter to develop a typology or taxonomy of these terms. Hence this introduction merely treats the concepts of knowledge and science in a very general, "common sense" way. Moreover, I make the simplistic assumption that we can treat knowledge and science synonymously.

In this chapter, there are three case studies involving the governance of knowledge: (1) Troy Duster's analysis of the norms currently employed by the scientific community in using "the concept of race," (2) Kevin Jones' study of how the British government responded to the problem of risk and uncertainty involving Bovine Spongiform Enceph-alopathy—or what is commonly known as "mad cow disease," and (3) Javier Lezaun's case study of some of the governance issues raised by the debate about genetically-modified foods.

Though these three cases raise many issues about the governance of science and knowledge, this introduction will focus on only three: (1) What are some of the conceptual issues which we should con-sider when we think about studying how science and knowledge are governed? (2) What relationship is there between the availability of information and thoughtful policy-making? (3) How do the fragmenta-tion of information and the specialization of research place constraints on the ability of actors to develop coherent policies about the world in which we are embedded?

Perspectives on Defining the Parameters of Governance

For many years, sociologists of science frequently wrote about the way in which science was governed by the scientific community (Merton 1973; Polanyi 1958). But as the following essays by Duster. Jones, and Lezaun, as well as others in this volume, demonstrate, contemporary scientific knowledge has enormous social consequences, and scientific research is enormously expensive. Because the governance of science has become a very complex and salient subject for our society, numer-ous actors other than scientists have become involved in the efforts to govern science.

As we attempt in many essays in this volume to understand the complexity of the governance of science, perhaps we can gain some insight into how to study the subject by borrowing from the literature on the governance of economic sectors. When we approach the problem this way, we quickly note that in most every sector there are common actors (e.g., producers, processors, and consumers). For example, in the steel and aluminum industries there are producers of ore, actors who process ore, and there are consumers or users of what has been processed. Similarly, in the scientific community, there are those who produce knowledge, those who process it into more hybrid areas of knowledge, as well as those who apply knowledge to develop some product (Hage and Hollingsworth 2000). Those who are processors of knowledge for one purpose may also be consumers for another. Each of these three groups is very heterogeneous in nature.

In most any sector of a society there are decisions about what to produce, who may engage in production, how much to produce, how to control the quality of the product, how the product is to be financed or funded, where the production will occur, how the production will be organized. Each of these problems must also be confronted in the world of science. For example, over the years various societies have faced the problem of what to produce: whether or not to do research on weapons of mass destruction, recombinant DNA, stem cell research, etc. There are the problems of regulating who may engage in scientific research (e.g., what are the minimum qualifications of those who are permitted to do research) and of regulating the quality of research. There are the problems of how much knowledge to produce. Some societies address this issue by limiting the number of people who may enter particular occupations. Finally, there are the problems of how to finance the production of knowledge and to determine who may consume the products produced by the scientific community. Many of these problems are addressed in one way or another in the three case studies that follow.

This immediately leads to one of the most important issues in studying the governance of any sector, and that is how to coordinate relations among actors in the same group (e.g., producers, processors, and consumers) and among actors across the different groups. This is also a theme that cuts across the following three case studies.

There are a variety of mechanisms for the governance or coordination of actors: the state, the market, corporate hierarchies, networks, and associations. Historically, associations (e.g., scientific societies, national academies) have been very much involved in governing or coordinating science. Moreover, networks have long been important as a governance mechanism. In some sense, much of the work of Thomas Kuhn (1962) and Michael Polanyi (1958) focuses on how relations among scientists were coordinated by networks. In more recent years, as scientists have attempted to sell the production of knowledge by means of such instruments as patents and licensing agreements, we observe the increasing importance of the market as a mechanism for the coordination of knowledge and science. Laboratories of large corporations such as Siemens, Philips, IBM, Monsanto, Du Pont, Merck, Bayer, and Hoechst are simply a few examples of how scientific activity is presently being coordinated by corporate hierarchies. This is a theme raised in Lezaun's paper, which follows.

There are at least two other important issues in addition to who the actors are, what problems are confronted and how relationships are coordinated. One is the relative power of actors. In studying the governance of any sector, it is extremely important that we be sensitive to the relative power of actors, and this is an issue which emerges from each of the following papers. Finally, there is the issue of the societal level at which governance occurs. This has become one of the most complicated issues involved in the governance of any sector. We are increasingly nested in a world in which decisions are made at multiple levels of reality at most any point in time. Decisions are made at the global level, the transnational regional level (e.g., the European Union), the nation-state, the sub-national regional and local levels. The papers by Duster, Jones, and Lezaun focus on how actors confront critical issues about the governance of knowledge at all of these different levels. The scholarly community needs to do much more theoretical and empirical work before we can properly understand how governance operates at all of these levels simultaneously. This is an important issue in two respects. We need to study governance at all of these multiple levels if we are to understand how the production of science and knowledge are presently coordinated. And from a normative point of view, we must attempt to understand the governance of knowledge and science at multiple levels in order to make more effective and informed choices with regard to the production of knowledge.

The Distinction between Having Information and Having Understanding

Never in history has there been such a broad dissemination of information about the world in which we are embedded. The amount of information on the Internet, which can quickly be accessed, is staggering. Indeed, there is an increasing democratization in the amount of information which is available to individuals throughout the world. And yet, with the increasing complexity of the world, even the most educated people of our time have a decreasing comprehension of the totality of what is going on in their environment—certainly compared to the most educated individuals of the late eighteenth century. Our scientific communities are ever challenging us with increasing amounts of new information. However, our understanding of the significance of the information and our ability to engage in coherent and effective policy making leaves a great deal to be desired. This is a major theme of the papers by Jones and Lezaun that follow. Technical and scientific developments are taking place at an accelerating pace, but as Everett Mendelsohn (1999: 90) has observed, social and moral disclosure about these developments "seems scattered, unfocused, and at times quixotic." In short there is a lack of disciplined analysis and understanding of the information which bombards us on a daily basis. Technical knowledge in the scientific community has far outstripped our capacity to develop effective strategies for regulating the consequences of new trends in science, and this is one of the great challenges of the time in which we are living. This is a strong message of the following three case studies.

Problems Resulting from the Fragmentation and Specialization of Research

Most observers would agree that the world's scientific community is performing extraordinarily well. On a decade by decade basis, there are more discoveries than ever before. But if the prime purpose of science is to learn as much as possible about the physical, biological, and social worlds and to make informed choices, the specialization and fragmentation of knowledge and the resulting breakdown in communication among scientists should become matters of serious concern.

Most social systems try to develop more effective integrative mechanisms as specialization and fragmentation occur. In the social system of science, theory has had the role of coordinating the community of scholars, of integrating new knowledge and guiding new research. Because so much knowledge is accumulating, however, the integrating mechanism of theory is constantly being strained. "There is a constant race between theory and research both in predicting new findings in advance and to find data that will bring about modification in theory" (Storer 1973: 146). As research findings proliferate at an accelerating rate it becomes increasingly difficult for our societies to integrate effectively new bodies of knowledge, and more and more investigators tend to work without regard to theory. With this growing lack of involvement in theoretical concerns, policy-makers and scientists tend to become ever more intellectually fragmented and to lack coherence in relating new information to solving the problems brought about by technical advances in their environment.

The increasing specialization and fragmentation of knowledge is leading to an authoritarian mentality in modern societies. On the one hand, our citizens have never had access to so much information, but at the same time our citizens fail to understand the world around them, leading them to accept the judgments of authorities in specialized fields—even though authorities lack comprehension of how specialized knowledge is related to other areas of high relevance. The only way of knowing something outside of one's specialty is to accept the word of an authority in a specialized field. Slowly, this tends to lead to an authoritarian political culture.

If our societies move too far in this direction, science may well begin to retrogress. Ideally, science should evolve in the direction of reducing the chaos of knowledge and perceived disunity. Some of the greatest achievements in science have been made by those master synthesizers who were able to reduce complex and wide-ranging phenomena into relatively simple, comparatively easy to understand relationships. But we are now tending to retrogress toward a condition whereby the chaos of reality is so complex that no group of individuals can comprehend important problems—much less propose solutions. As a result, one of the great challenges of our time is to devise a means of coordinating and governing of science so that we can relate our technology and information to advancing the welfare of our citizens. The following three papers are fascinating case studies of how our scientific communities are so fragmented and specialized that they are unable to develop the social and moral perspectives to devise and implement effective regulatory and governance strategies for our world.

References

Campbell, John. J. Rogers Hollingsworth, and Leon Lindberg. eds. (1991) *The Governance of the American Economy.* New York: Cambridge University Press.

Edelman, Gerald. 12 February (1998) Interview with Rogers Hollingsworth at the Neurosciences Institute, La Jolla, Cal.

Greenspan, Ralph. 7 January, 13 March (1998) Interviews with Rogers Hollingsworth at the Neurosciences Institute, La Jolla, Cal.

Hollingsworth, J. Rogers, and Robert Boyer, eds. (1997) *Contemporary Capitalism: The Embeddedness of Institutions.* New York: Cambridge University Press.

Hollingsworth, J. Rogers, Philippe Schmitter, and Wolfgang Streeck (1994) *Governing Capitalist Economies: Performance and control of economic sectors.* New York: Oxford University Press.

Hage, Jerald, and J. Rogers Hollingsworth (2000) "Idea innovation networks: A strategy for integrating organizational and institutional analysis," *Organization Studies* 21: 971–1004.

Kuhn, Thomas (1962) *The Structure of Scientific Revolutions.* Chicago: University of Chicago Press.

Mendelsohn, Everett (1999) "Is Public Policy Lagging Behind the Science?" in *Office of Health Economics, Genomics, Healthcare, and Public Policy.* London: Office of Health Economics.

Merton, Robert (1973) "The Matthew Effect in Science," in *The Sociology of Science: Theoretical and empirical investigations.* Chicago: University of Chicago Press. 452–3.

Polanyi, Michael (1958) *Personal Knowledge: Towards a post-critical philosophy.* Chicago: University of Chicago Press.

Storer, Norman W. (1973) *The Sociology of Science.* Chicago: University of Chicago Press.

7

Feedback Loops in the Politics of Knowledge Production*

Troy Duster

A consortium of leading scientists across the disciplines from biology to physical anthropology issued a "Revised UNESCO Statement on Race" in 1995—a definitive declaration that summarizes eleven central issues, and concludes that in terms of "scientific" discourse, there is no such thing as a "race" that has any scientific utility:

> the same scientific groups that developed the biological concept over the last century have now concluded that its use for characterizing human populations is so flawed that it is no longer a scientifically valid concept. In fact, the statement makes clear that the biological concept of race as applied to humans has no legitimate place in biological science (Katz 1995: 4,5).

Note that the statement is not only about the utility of the concept of race for biological science. Rather, it asks in its title, "Is race a legitimate concern for science?" and in the quotation above, states that the concept "is so flawed that it is no longer a scientifically valid concept." For more than two centuries, the intermingling of scientific and common-sense thinking about race has produced remarkable trafficking back and forth between scientists and the laity, confusing for both laypersons and scientists about the salience of race as a stratifying practice (itself worthy of scientific investigation) *versus* race as a socially de-contextualized biologically accurate and meaningful taxonomy. The current decade is no exception. In the rush to purge common-sense

*An expanded version of this article appears as "Buried Alive: The Concept of Race in Science," in *Genetics, Race and Culture*, in Alan Goodman, Deborah Heath, and Susan Lindee (eds.). Berkeley, Cal: University of California Press (2003).

thinking of groundless belief systems about the biological basis of racial classifications, the current leadership of scientific communities has over-stated the simplicity of very complex interactive feedback loops between biology and culture and social stratification.

I will demonstrate how and why "purging science of race"—where race and ethnic classifications are embedded in the routine collection and analysis of data (from oncology to epidemiology, from hematology to social anthropology, from genetics to sociology)—is neither practicable, possible, nor even desirable. Rather, our task should be to recognize, engage, and clarify the complexity of the interaction between *any* taxonomics of race and biological, neurophysiological, social, and health outcomes. Whether or not race is a legitimate concept for scientific inquiry depends upon the designation of the unit of analysis of "race," and will in turn be related to the purposes for which the concept is deployed. This may seem heretical at the outset, but my rescues an important role for examining the purpose of an investigation to legitimize the analytic utility of the concept of race.

My strategy will be threefold. First, I will summarize an emerging clinical genetics problem from recent blood studies that is now forcing scientific medicine to reconsider the practical or efficacious meaning of race when it comes to blood transfusions. Second, I will turn to recent attempts to identify individuals from ethnic and racial populations through the use of the new technologies of molecular genetics. Here it is vital to note the emphasis upon the *practical* applications of these technologies, from their uses in forensics (the exclusion or probable identification of suspects in criminal investigations) to pharmacogenomics—a field that explicitly deploys the concept of "race" in the attempt to focus the delivery of pharmaceuticals to populations so designated, and does not bother to place quotation marks around the concept. Third, I will briefly point to the possible, even likely interaction between racial or ethnic identity, nutritional intake, and biochemical manifestation of disease states, most notably, cancer and heart disease. Finally, I will suggest a way to address, even resolve the confusing and contradictory messages about "race" from the biological sciences and their applied satellites.

I will conclude with some remarks about how anthropologists (and others working on aggregate data on selected populations designated by "race") should try to advance our understanding of *how* "race" is always going to be a complex interplay of social and biological realities with ideology and myth.

Context and Content for Feedback Loops:
Setting the Empirical Problem

By the mid 1970s, it had become abundantly clear that there is more genetic variation within the most current common socially used categories of race than between these categories (Polednak 1989; Bittles and Roberts 1992; Chapman 1993; Shipman 1994). The consensus is a recent development. For example, in the early part of the twentieth century, scientists in several countries tried to link up a study of the major blood groups in the ABO system to racial and ethnic groups.[1] They had learned that blood type B was more common in certain ethnic and racial groups—which some believed to be more inclined to criminality and mental illness (Gundel 1926; Schusterov 1927). They kept running up against a brick wall because there was nothing in the ABO system that could predict behavior. While *that* strategy ended a full half-century ago, there is a contemporary arena in which hematology, the study of blood, has had to resuscitate a concern with "race."

In the United States there has been an increasing awareness developed over the last two decades of the problem that blood from Americans of European ancestry (read mainly white) tends to contain a greater number of antigens than blood from Americans of African or Asian ancestry (Vichinsky et al. 1990). This means that there is a greater chance for hemolytic reactions for blacks and Asians receiving blood from whites, but a lower risk for whites receiving blood from Asians or blacks. Here we come to a fascinating intersection between the biological and social sciences. In the United States, not only do whites comprise approximately 80 percent of the population, proportionally fewer blacks and fewer Asian Americans donate blood than do whites. This social fact has some biological consequences, which in turn have some social consequences.

Approximately four hundred red blood cell group antigens have been identified. The antigens have been classed into a number of fairly well-defined systems: the most well known are the ABO and Rh systems, but there are other systems such as P, Lewis, MN, and Kell (standard hematology texts note ten systems, including ABO and Rh). The clinical significance of blood groups is that in the case of a blood transfusion, individuals who lack a particular blood-group antigen may produce antibodies reacting with that antigen m the transfused blood. This immune response to alloantigens (non-self antigens) may produce hemolytic reactions, the most serious being complete hemolysis (destruction of all red blood cells), which can be life-threatening. Once generated, the capacity to respond to a particular antigen is more or less permanent because the immune system generates "memory cells" that can be activated by future exposures to the antigen. For those who have chronic conditions that require routine blood transfusion, this aspect of the immune response is critical because it increases the likelihood of future transfusion incompatibility. The clinical goal is to minimize immune responses to antigens in transfused blood, in part because a crisis (such as trauma surgery) may require transfusion of whatever blood is available, regardless of its antigen composition.

Most blood banks only test for ABO and Rh—the most common systems. Testing for the other systems is considered inefficient and will increase the cost of blood. It is essential to minimize the antibodies against blood group antigens for everyone. However, the way in which blood typing is done puts members of racial and ethnic minorities at greater risk for the negative consequences of frequent transfusions. The term *phenotypically matched blood* basically means that it is possible to use the social appearances of race as a rough approximation (of likely antigens) to screen to minimize antibodies (along with ABO and Rh).

Transfusion therapy for sickle cell anemia is limited by the development of antibodies to foreign red cells (Vichinsky et al. 1990). In one important study, the researchers evaluated the frequency and risk factors associated with such alloimmunization, and obtained the transfusion history, red-cell phenotype, and development of alloantibodies in 107 black patients with sickle-cell anemia who received transfusions. They then compared the results with those from similar studies in 51 black patients with sickle-cell anemia who had not received transfusions and in 19 non-black patients who received transfusions for their forms of chronic anemia.

> We assessed the effect that racial differences might have in the frequency of alloimmunization by comparing the red-cell phenotypes of patients and blood-bank donors (n=200, 90 percent white). Although they received transfusions less frequently, 30 percent of the patients with sickle cell anemia became alloimmunized, in contrast to 5 percent of the comparison-group patients with other forms of anemia (P less than 0.001). Of the 32 alloimmunized patients with sickle cell anemia, 17 had multiple antibodies and 14 had delayed transfusion reactions. Antibodies against the K, E, C, and Jkb antigens accounted for 82 percent of the alloantibodies.

They then go on to conclude:

> These differences are most likely racial. We conclude that alloimmu-
> nization is a common, clinically serious problem in sickle cell anemia
> and that it is partly due to racial differences between the blood-donor
> and recipient populations (Vichinsky et al. 1990).

True enough, these "racial differences" may not be "race" in any essen-
tialist conception, but that is precisely the point. A full eighty years ago,
Hirschfeld and Hirschfeld (1919: 675) posited that when introducing
the blood of one species into that of the same species "those antigen
properties which are common to the giver and receiver of blood can
not give rise to antibodies, since they are not felt as foreign by the
immunized animal." While the Hirschfelds were talking about dogs,
they were drawing a straight line towards humans, human classification,
and racial taxonomy:

> If we inject into dogs the blood of other dogs it is in many cases pos-
> sible to produce antibodies. By means of these antibodies we have
> been able to show that there are in dogs two antigen types. These
> antigen types, which we recognize by means of the iso-antibodies,
> may designate two biochemical races. (pp. 675–76)

Using this hypothesis, they went on to perform the first systematic
and comprehensive serological study of a variety of ethnic and racial
groups. As I indicated above, their classification system did not survive the test of time, but "a way of thinking" persists (Marks 1995).
Moreover, with the data reported in the Vichinsky study (given that
the increased blood donations from blacks is a key policy goal for
those suffering from a relatively common genetic disease—sickle-cell
anemia), the resuscitation of "race" through blood antigen theorizing
and empirical research means that the concept is very much still with
us in clinical medicine. That persons who are phenotypically "white"
can and do have sickle-cell anemia complicates any essentialized racial
theorizing to be sure—but for the purposes of further action (blood
donation requests and transfusion direction), racial phenotyping as a
"short-hand" is back with us at the end of the twentieth century.

This provides a remarkably interesting intersection. While the full
range of analysts, commentators, and scientists—from post-modern
essayists to molecular geneticists to social anthropologists—have been
busily pronouncing "the death of race," for practical clinical purposes

the concept is resurrected in the conflation of blood donation frequencies by "race." I am not merely trying to resurrect "race" as a social construct (with no biological meaning)—no more than I am trying to resurrect "race" as a biological construct with no social meaning. Rather, I am arguing that when "race" is used as a stratifying practice (which can be apprehended empirically and systematically) there is often a reciprocal interplay of a biological outcomes that makes it impossible to completely disentangle the biological from the social. While that may be obvious to some, it is completely alien to others, and some of those "others" are key players in current debates about the biology of race.

In late September 1996, Tuskegee University hosted a conference on the Human Genome Project, with specific reference to the Project's relevance to the subject of race (Smith and Sapp 1997). In attendance was Luca Cavalli-Sforza, a pre-eminent population geneticist from Stanford University and perhaps the leading figure behind the Human Genome Diversity Project.[2] Cavalli-Sforza had appeared on the cover of *Time* magazine a few years earlier, as something of a hero to the forces that were attacking the genetic determinism in *The Bell Curve*.[3] At this conference, he repeated what he had said in the *Time* article: "One important conclusion of population genetics is that races do not exist" (ibid. : 53).

> If you take differences between two random individuals of the same population, they are about 85% of the differences you would find if you take two individuals at random from the whole world. This means two things: (1) The differences between individuals are the bulk of the variation; (2) the differences among populations, races, continents are very small—the latter are only the rest, 15%, about six times less than that between two random individuals of one perhaps very small population (85%). Between you and your town grocer there is on average a variation which is almost as large as that between you and a random individual of the whole world. This person could be from Africa, China, or an Australian aborigine. (p. 55).

Cavalli-Sforza is speaking here as a population geneticist, and in that limited frame of what is important and different about us as humans, he may be empirically correct. But humans give meaning to differences. At a particular historical moment, to tell this to an Albanian in Kosovo, a Hutu among the Tutsi, a Zulu among the Boers, or to a German Jew among the Nazis, may be as convincing, for the purposes of further action, as telling it to an audience of mainly African Americans at Tuskegee University.[4] Indeed, David Botstein, speaking later in a keynote address, had this to say about *The Bell Curve*:

> So from a scientific point of view, this whole business of *The Bell Curve*, atrocious though the claims may be, is nonsense and is not to be taken seriously. People keep asking me why l do not rebut *The Bell Curve*. The answer is because it is so stupid that it is not rebuttable. You have to remember that the Nazis who exterminated most of my immediate family did that on a genetic basis, but it was false. Geneticists in Germany knew that it was false. The danger is not from the truth, the danger is from the falsehood. (p. 212)

David Botstein is also a pre-eminent molecular geneticist at the vanguard of his field. In this statement, he takes the position that if people just understood the genetic truth, that would be sufficient and even corrective of racist thinking and action.[5] Even though people may someday come to understand that they are basically similar at the level of the DNA, RNA, immunological or kinds of blood systems—it is the language group, kinship, religion, region, and race that are still far more likely to generate their pledge of allegiance.

The American Anthropological Association Statement on "Race"

In May 1998, the American Anthropological Association issued its own statement on "race" (1998). It attempts to address the myths and misconceptions, and in so doing takes a "corrective" stance towards the folk beliefs about race. The statement strongly states the position that "physical variations in the human species have no meaning except the social ones that humans put on them." But in casting "the problem" in this fashion, it gives the impression that the biological meanings that scientists attribute to race are biological facts, while the social meanings that laypersons give to race are first either errors or mere artificial social constructions, and second not themselves capable of feedback loops into the biochemical, neurophysiological, and cellular aspects of our bodies that in turn, can be studied, scientifically. The statement of the Anthropological Association is consistent with that of the UNESCO statement on race. However, by formulating the matter so that it is *only* the social meanings that humans provide implies that mere lay notions of race provide a rationale for domination, but have no other utility.

There is profound misunderstanding of the implications of a "social constructivist" notion of social phenomena. How humans identify themselves, whether in religious or ethnic or racial or aesthetic terms, influences their subsequent behavior. Places of worship are socially

153

constructed with human variations of meaning and interpretation and use very much in mind. Whether a cathedral or mosque, a synagogue or Shinto temple, those "constructions" are no less "real" because one has accounted for and documented the social forces at play that resulted in such a wide variety of "socially constructed" places of worship. "Race" as social construction can and does have a substantial effect on how people behave. One important arena for further scientific exploration and investigation is the feedback between that behavior and the biological functioning of the body. It is now appropriate to restate the well-known social analytic aphorism of W.I. Thomas, but to refocus it on human taxonomics of other humans: *If humans define situations as real, they can and often do have real biological and social consequences.*

Explicating the Conflation of Crime, Genetics, and Race

If "race" is a concept with no scientific utility, what are we to make of a series of articles that have appeared in the scientific literature over the last seven years, looking for genetic markers of population groups that coincide with common-sense, lay renditions of ethnic and racial phenotypes? It is the forensic applications that have generated much of this interest. Devlin and Risch (1992a) published an article on "Ethnic differentiation at VNTR loci, with specific reference to forensic applications"—a research report that appeared prominently in the *American Journal of Human Genetics.*

> The presence of null alleles leads to a large excess of single-band phenotypes for blacks at D17S79.... This phenomenon is less important for the Caucasian and Hispanic populations, which have fewer alleles with a small number of repeats (p. 540)

> ... it appears that the FBI's data base is representative of the Caucasian population. Results for the Hispanic ethnic groups, for the D17S79 locus, again suggest that the data bases are derived from nearly identical populations, when both similarities and expected biases are considered. . . . For the allele frequency distributions derived from the black population, there may be small differences in the populations from which the data bases are derived, as the expected bias is .05. (p. 546)

When researchers try to make probabilistic statements about which group a person belongs to, they look at variation at several different locations in the DNA—usually from three to seven loci. For any particular locus, there is an examination of the frequency of that allele at that locus, and for that population. In other words, what is being assessed is the frequency of genetic variation at a particular spot in the DNA in each population.

Occasionally, these researchers find a locus where one of the populations being observed and measured has, say, alleles H, I, and J, and another population has alleles H, I, and K. For example, we know that there are alleles that are found primarily among sub-populations of native. Americans. When comparing a group of native Americans with a group of Finnish people, one might find a single allele that was present in some natives but in no Finns (or its at such a low frequency in the Finns that it is rarely, if ever seen). However, it is important to note and reiterate again and again that this does not mean that all natives, even in this sub-population, will have that allele.[6] Indeed, it is inevitable that some will have a different set of alleles, and that many of them will be the same alleles as some of the Finns. Also, if comparing natives from Arizona to North American Caucasians from Arizona, we would probably find a low level of the "native allele" in the so-called Caucasians, because there has been "interbreeding." Which leads to the next point.

It is possible to make arbitrary groupings of populations (geographic, linguistic, self-identified by faith, identified by others by physiognamy, etc.) and still find statistically significant allelic variations between those groupings. For example, we could simply pick all of the people in Chicago, and all in Los Angeles, and find statistically significant differences in allele frequency at *some* loci. Of course, at many loci, even most loci, we would not find statistically significant differences. When researchers claim to be able to assign people to groups based on allele frequency at a certain number of loci, they have chosen loci that show differences between the groups they are trying to distinguish. The work of Devlin and Risch (1992a, 1992b), Evett et al. (1993, 1996) and others suggest that there are only about 10 percent of sites in the DNA that are "useful" for making distinctions. This means that at the other 90 percent of the sites, the allele frequencies do not vary between groups such as "Afro-Carri bean people in England" and "Scottish people in England." But it does not follow that because we can not find a single site where allele frequency matches some phenotype that we are trying to identify (for forensic purposes, we should be reminded), that there are not several (four, six, seven) that will not be effective, for the purposes of aiding the FBI, Scotland yard, or the criminal justice systems around the globe in highly probabilistic statements about suspects, and the likely ethnic, racial, or cultural populations from which they can be identified—statistically.

An article in the 8 July 1995 issue of the *New Scientist* entitled "Genes in black and white" details some extraordinary claims made about what it is possible to learn about socially defined categories of race from reviewing information gathered using new molecular genetic technology (Vines 1995):

> In 1993, a British forensic scientist published what is perhaps the first DNA test explicitly acknowledged to provide "intelligence information" along "ethnic" lines for "investigators of unsolved crimes." Ian Evett, now at the Home Office's forensic science laboratory in Birmingham, and his colleagues in the Metropolitan Police, claim that their DNA test can distinguish between "Caucasians" and "Afro-Caribbeans" in nearly 85 percent of the cases. . . . Evett's work, published in the *Journal of Forensic Science Society,* draws on apparent genetic differences in three sections of human DNA. Like most stretches of human DNA used for forensic typing, each of these three regions differs widely from person to person, irrespective of race. But by looking at all three, say the researchers, it is possible to estimate the probability that someone belongs to a particular racial group.

The implications of this for determining, for legal purposes, who is and is not "officially" a member of some racial or ethnic category are profound.

Two years after the publication of the UNESCO statement purportedly burying the concept of "race" for the purposes of scientific inquiry and analysis, and during the same time period that the American Anthropological Association was deliberating and generating a parallel statement, an article appeared in the *American Journal of Human Generics,* authored by Ian Evett and his associates, summarized thusly:

> Before the introduction of a four-locus multiplex short-tandem-repeat (STR) system into casework, an extensive series of tests were carried out to determine robust procedures for assessing the evidential value of a match between crime and suspect samples. Twelve databases were analyzed from the three main ethnic groups encountered in casework in the United Kingdom; Caucasians, Afro-Caribbeans, and Asians from the Indian subcontinent. Independence tests resulted in a number of significant results, and the impact that these might have on forensic casework was investigated. It is demonstrated that previously published methods provide a similar procedure for correcting allele frequencies —and that this leads to conservative casework estimates of evidential value. (Evett et al. 1996: 398)

These new technologies have some not-so-hidden potential to be used for a variety of forensic purposes in the development and "authentication" of typologies of human ethnicity and race. A contemporary update of an old idea of the idea of deciding upon "degree of whiteness" or "degree of nativeness" is possibly upon us, anew, with the aid of molecular genetics. Vines (1995) describes the Allotment Act of 1887, denying land rights to those native Americans who were "less than half-blood." The U.S. government still requires American Indians to produce "Certificates with Degree of Indian Blood" in order to qualify for a number of entitlements, including being able to have one's art so labeled. The Indian Arts and Crafts Act of 1990 made it a crime to identify oneself as a Native American when selling artwork without federal certification authorizing one to make the legitimate claim that one was, indeed, an authentic ("one-quarter blood" even in 1990s) American Indian. As noted above, it is not art, but law and forensics that ultimately will impel the genetic technologies to be employed in behalf of attempts to identify who is "authentically" in one category or another. Geneticists in Ottawa, Canada have been trying to set up a system "to distinguish between 'Caucasian Americans' and 'Native Americans' on the basis of a variable DNA region used in DNA fingerprinting" (Vines 1995: 37). For practical purposes, the issue of the authentication of persons' membership in a group (racial/ethnic/cultural) can be brought to the level of DNA analysis. The effectiveness of testing and screening for genetic disorders in risk populations that are ethnically and racially designated poses a related set of vexing concerns for the "separation" of the biological and cultural taxonomies of race.

Genetic Testing and Genetic Screening

When social groupings with a strong endogamous tradition (such as ethnic or racial groups) intermarry for centuries, they are at higher risk for pairing recessive genes and passing on a genetic disorder. In the United States, the best knowns of these clustered autosomal recessive disorders are Tay-Sachs disease, beta-thalassemia, sickle-cell anemia, and cystic fibrosis. For Tay-Sachs, concentrated primarily among Ashkenazi Jews of northern and eastern European ancestry, about one in thirty is a carrier, and approximately one in every 3,000 newborns will have the disorder. For cystic fibrosis, about one in thirty Americans of European descent is a carrier, with a similar incidence rate.[7] In contrast, approximately one in every 12 American blacks is a carrier for sickle-cell anemia and one in every 625 black newborns will have the disorder. Irish and northern Europeans are at greater risk for phenylketonuria. In the United States, one in 60 Caucasians is a carrier, and about one in every 12,000 newborn Caucasians is affected (see Tables 7.1 and 7.2).

When both parents are carriers of the autosomal recessive gene, the probability that each live birth will be affected by the disorder is 25 percent. However, being a carrier, or passing on the gene so that one's offspring is also a carrier, typically poses no more of a health threat than carrying a recessive gene for a different eye color. That is, carrier status typically poses no health threat at all. The health rationale behind carrier screening is to inform prospective parents about their chances of having a child with a genetic disorder.

In the United States, the two most widespread genetic screening programs for carriers have been for Jews of northern European descent (Tay-Sachs) and for Americans of western African descent (sickle-cell anemia) From 1972 to 1985, there was widespread prenatal screening for both disorders, and by 1988, newborn screening for sickle-cell anemia had become common (Duster 1990). It is the autosomal recessive disorders, located in risk populations that coincide with ethnicity and race, that are of special interest as we turn to address genetic screening for populations that are at greatest risk for a disorder.

It is important to distinguish between a genetic screen and a genetic test. A genetic test is done when there is reason to believe that a particular individual is at high risk for having a genetic disorder, or for being a carrier of a gene (recessive) for a disorder. So for example, a sibling of someone who has been diagnosed with Huntington's (a late-onset neurological disorder) would be a candidate for a genetic test for that disorder. A genetic screen, on the other hand, is used for a *population* that is at higher risk for a genetic disorder. Thus, with the risk figures cited above, Ashkenazi Jews were the subjects of genetic *screening* for Tay-Sachs.

As with most of the genetic disorders mentioned above, the incidence of cystic fibrosis varies remarkably with different groups. For the purposes of my argument, what is most striking is that *the sensitivity of the current genetic test for cystic fibrosis varies remarkably with different groups*. For example, the sensitivity of the test is 97 percent for Ashkenazi Jews, but only 30 percent for Asian Americans. The National Institutes of Health convened a Consensus Development Conference on Genetic Testing for Cystic Fibrosis, 14–16 April 1997. As a result of that meeting, a statement was issued declaring that cystic fibrosis testing should be made available to all couples planning a pregnancy. Yet, with such a low sensitivity for some groups, there are going to be

Table 7.1. Selected High Incidence of Genetics Disorders.

Condition	Estimated Incidence*
cleft lip/palate	1 in 675 individuals
club foot	1 in 2,500 Caucasians
diabetes	1 in 80 individuals
Down syndrome	1 in 1,050 individual
hemophilia	1 in 10,000 males
Huntington's disease	1 in 2,500 individuals
Duchenne muscular dystrophy	1 in 7,000 males
phenlyketonuria (PKU)	1 in 12,000 Caucasians
Rh incompatibility	1–2 in 100 individuals
sickle cell anemia	1 in 625 African Americans
Tay-Sachs disease	1 in 3,000 Ashkenazi Jews
beta-thalassemia (Cooley's anemia)	1 in 2,500 Mediterranean people

*The above figures vary according to the ethnic background. For example, Rh incompatibility is much lower among those with Asian ancestry. Phenylketonuria is rare in those of black or Ashkenazi Jewish ancestry (most Jewish people in the United States are of Ashkenazi descent). (From L. Burhansstipanov, S. Giarratano, K. Koser and J. Mongoven, "Prevention of Genetic and Birth Disorders," Sacramento: California State Department of Education, 1987: 6–7.)

Table 7.2. Ethnicities/Groups Primarily Affected by Disorders (USA).

Condition	Ethnicities Primarily Affected
Duchenne muscular dystrophy	northeastern British
adult lactase deficiency	African-Americans
cleft lip/palate	North American Indians, Japanese
cystic fibrosis	northern Europeans
familial Mediterranean fever	Armenians
phenlyketonuria (PKU)	Caucasians (especially Irish)
sickle-cell anemia	African-Americans
alpha-thalassemia	Chinese, southeast Asians
beta-thalassemia	Mediterraneans
ppina bifida/anencephaly	Caucasians (especially Welsh and Irish)
Tay-Sachs disease	Ashkenazi Jew's (origins in central and eastern Europe)

(From L. Burhansstipanov, S. Giarratano, K. Koser and J. Mongoven Prevention of Genetic and Birth Disorders, Sacramento: California State Department of Education, 1987: 6–7.)

Table 7.3. Ethnic/Group Variation with Incidence of Cystic Fibrosis, with Sensitivity to DeltaF508 Test.

Group	Incidence	Carrier Frequency	% Delta F508	Sensitivity
Caucasians	1:3,300	1.29	70	90
Ashkenazi Jew's		1:29	30	97
Native Americans	1:3,950		0	94
Hispanics	1:8,500	1:46	46	57
African-Americans	1:15,300	1:63	48	75
Asian-Americans	1:32,100	1:90	30	30

many false positives.[8] Notice that the sensitivity among Latinos is only 57 percent, despite relatively high incidence.

The total incidence of cystic fibrosis among native Americans is only 1 in 11,200—but for the Zuni, the rate is 1 in 1,580; seven times higher. Moreover we now know that there is a particular allele for cystic fibrosis peculiar to Zuni Indians. It is inevitable that an allele associated with the drinking patterns of certain social groups can be "marked"—even if it turns out to be a spurious relationship. Given the nature of the quest, it is likely that research scientists are likely to find an allele associated with certain native American tribes—and their generally higher rates of alcoholism. We already know that Asians are more likely to have the allele that produces flushing from alcohol. If we contrast Ashkenazi Jews with Zuni Indians in their respective capacities to direct or divert genetic testing based upon ethnicity, race, or relative social power in the medical profession, we would find some interesting patterns. Ashkenazi Jewish women have begun to strongly protest their identification as being at higher risk for breast cancer. The Zuni have yet to be heard from in a parallel manner regarding their identification with cystic fibrosis.

Pharmacogenomics as the Harbinger of Germline Intervention

In the last few years, the field of pharmacogenomics has begun to develop around the delivery of pharmaceuticals to population-specific groups. The new pharmacogenomics asserts unequivocally that there are racial differences in the way "different races" respond to certain drugs. Writing in *Science,* Evans and Rolling claim that "all pharmacogenetic polymorphisms studied to date differ in frequency among ethnic and racial groups" (Evans and Rolling 1999). Whether or not this is based on thoughtfully controlled subject populations, this helps explain the recent announcement that the Food and Drug Administration has just given a provisional green light for a pharmaceutical company (NitroMed) to proceed to try to market "the first ethnic drug," BiDil. It is a drug for heart disease specifically designed for the African American population. Blacks are reported clinically to have higher blood pressure rates than whites, and are twice as likely as whites to suffer heart failure. This opened the door to biotechnology companies seeking to develop and market drugs to be ethnically and even "racially" specific. In early 2001. NitroMed developed a drug designed specifically for African Americans.

It is not much of a conceptual leap from a pharmaceutical designed for a particular "population group" and a germline intervention designed for such a group. Yet, since economic profit wall drive the engine of biotechnology (unashamedly, proudly pronounced as the *sine qua non* of good business in a capitalist society), a germline intervention for the Zuni is not in the cards—or perhaps, more realistically, not in the profit margin.

The Interaction between Race as Identity, Nutrient Consumption, and Health

The scientific literature on the rates of specific cancers in racially and ethnically designated populations is fairly well-developed. For example, Ashkenazic Jewish women are reported, clinically, to have higher rates of breast cancer than other groups (Richards et al. 1997: 1096). African-American men have almost double the rate of prostate cancer of white men in certain age groups, according to reports released by the National Cancer Institute (Ries et al. 2002). How might this be explained, using race not simply as an "outcome"—but as a factor that helps produce the outcome? Consider the possibility that certain forms of cancer may be a function of nutrition and diet. Groups with certain dietary patterns or restrictions might then be systematically (i.e., apprehensive scientifically) at greater risk for cancer. If members of a certain group identify themselves as say, Ashkenazic Jewish, and then have a diet that follows certain patterns, they might well routinely have rates of certain groups of cancers, at both lower and higher risks than groups with different dietary habits. African American males, for example, may, by identifying as African Americans, be more likely to eat a category of food ("soul food") that might systematically put them at higher risk for prostate cancer. With this formulation, I am "bringing the systematic study of race" back into the "scientific inquiry"—even though I am not going to the molecular level to attempt a reductionist account of "race as caused" at the level of the DNA.

Here is where the computer revolution enters the story: Up until very recently, we could not do much with these random variations in the DNA, called single nucleotide polymorphisms—SNPs. However, with the new computers, we can now put the DNA of several clusters of people on computer chips, and see what might be patterns in their DNA (Hamadeh and Afshari 2000).

It is now possible to do hundreds, even thousands of experiments in a few hours. This might prove to be a useful technology in the hunt for particular regions that might help explain some illnesses. For example, if we get a few hundred patients, all with prostate cancer—we can then look at their SNP profile using this chip technology, with the hope the hope of discerning some patterns to serve as markers to help locate candidate genes. A similar strategy is being deployed with the search for the genetic contributions to explanations of heart disease.

With these new SNPs on chips, we are likely to come up with new taxonomies that include new aggregations of people who share certain patterns in their DNA.

Even with strong epidemiological evidence that heart disease and hypertension among African Americans is strongly associated with such social factors as poverty, there has been a persistent attempt to pursue the scientific study of hypertension through a link to the genetics of race. Dark pigmentation is indeed associated with hypertension in America. Michael Klag et al. (1991) reported the results of a carefully controled study looking at the relationship between skin color and high blood pressure. He and his colleagues found that darker skin color is a good predictor of hypertension among blacks of low socioeconomic status, but not for blacks of any shade who are "well employed or better educated." The study further suggested that poor blacks with darker skin color experience greater hypertension "not for genetic reasons" but because darker skin color subjects them to greater discrimination, with consequently greater stress and psychological/medical consequences. Of course, from another way of looking at it, "darker skin color" is dark *mainly* for genetic reasons, so it is all a matter of how one chooses to direct theorizing about the location of causal arrows. When practicing physicians see "darker skin color," their diagnostic interpretation and their therapeutic recommendations are systematically affected. Schulman et al. (1999) recently published some research indicating that in clinical practice, physicians are likely to make systematically different recommendations for treatment of heart disorders, by race, even when patients present the same symptoms. Thus, when there is an analysis of outcome data such as "cause of death" by race, and researchers find that blacks have a higher incidence of death from heart failure—it would be easy to make an incorrect inference about causation and direction of the relationship between the variables.

When African Americans show up in clinical populations with higher rates of prostate cancer, and the prevailing paradigm for oncology is molecular genetics—there will be a social tendency for some genetic detective-work for the search for oncogenes (Fujimura 1996). Nancy Press et al. (1998) brought an anthropological lens to the matter that was discussed earlier, the report in the literature that Ashkenazic Jewish women are at higher risk, for certain breast cancers (Richards et al. 1997).

When the National Institutes of Health convened a panel to review its full portfolio of research on violence at the beginning of this decade, the panel concluded that the great bulk of studies focused on the individual, or smaller, units of analysis. Nonetheless, while accounting for only 12 percent of the population, blacks are convicted of over 60 percent of the homicides. When queried about this, the answer from research scientists working on neurological or bio-chemical level was that "we are engaged in basic science". What this meant, empirically, was that they regarded any larger units (beyond the individual) dealing with violence are simply "not science" but are "policy" matters. So, that is the justification for assigning the bulk of science funding to "basic processes" (Daniel S. Greenberg, "Hardly an ounce for prevention" *Washington Post* 22 March 1999). But how can the molecular geneticists and neural sciences have it both ways? If they want to lay scientific claim to be able to better explain the "basic processes of violence" behind the high rate of homicide in the African American community, all the while saying that there is nothing to the *scientific* classification of race, then by their own admission the study of violence at the molecular level, or at the level of neuro-transmission (serotonin levels remains the most popular theory) will have nothing to do with race as phenotypically reported in the FBI crime statistics.

By heading toward an unnecessarily binary, socially constructed fork in the road, by forcing ourselves to think that we must either choose between either "race as biological" (now out of favor) and "race as *merely* a social construction" we fall into an avoidable trap. A refurbished and updated insight from W. I. Thomas can help us. It is not an either/or proposition. Under some conditions, we need to conduct systematic investigation, guided by a body of theory, into the role of "race" (or ethnicity, or religion) as an organising force in social relations, and as a stratifying practice (Oliver and Shapiro 1995). Under other conditions, we will need to conduct systematic investigation, guided by a body of theory, into the role of the interaction of "race" (or ethnicity, or religion) however flawed as a biologically discrete and coherent taxonomic system, with feedback loops into the biological functioning of the human body; or with medical practice. The latter studies might include examination of the systematic administration of higher doses of x-rays to African Americans; the creation of genetic tests with high rates of sensitivity to some ethnic and racial groups, but low sensitivity to others; and the systematic treatment, or lack of it, with diagnostic and therapeutic interventions to "racialized" heart and cancer patients.

It is not difficult to understand why they persisted. Humans are symbol-bearing creatures that give meaning to their experiences and to their symbolic worlds. The UNESCO statement is ultimately about the problem of the difference between first-order constructs in science, versus second-order constructs. Some fifty years ago, Felix Kaufmann ([1944] 1958) made a crucial distinction that throws some light on the controversy. Kaufmann was not addressing whether or not there can be a science of race. Rather, he noted that there are different kinds of issues, methodologies, and theories that are generated by what could be called "first-order constructs" in the physical and natural sciences versus "second-order constructs." For the physical and natural sciences, the naming of objects for investigation and inquiry, for conceptualizing and finding empirical regularities, is in the hands of the scientists and their scientific peers. Thus, for example, the nomenclature for quarks or neurons, genes or chromosomes, nitrogen or sulfides, etc., all reside with the scientist in his/her role as the creator of first-order constructs.

This is quite different from the task of the observer, analyst, or scientist of human social behavior. This is because humans live in a pre-interpreted social world. They grow up, from infancy, in a world that has pre-assigned categories and names for those categories, which were in turn provided by fellow common-sense actors, no! by "scientists" (Schutz 1973). Their continual task is try to navigate, negotiate and make sense of that world. The task of the social scientist is therefore quite distinct from that of the natural scientist. While the latter can rely upon "first-order constructs," the former must construct a set of categories based upon the pre-interpreted world of common-sense actors. The central problem is that "race" is now, and has been since 1735,[9] both a first- and second-order construct. The following joke, making the rounds among African-American intellectuals and reported in Roediger (1994: 1), makes the point with deft humor; "I have noted," the joke laments, "that my research demonstrating that race is merely a social and ideological construction helps little in getting taxis to pick me up late at night."

This throws into a different light the matter of whether race can be studied scientifically. If we mean by that, is there a consensus among the natural scientists about race as a "first-order construct," then the answer since about 1970 is categorically "no." The UNESCO statement summarizes why this is so at every level that is significant to the biological functioning of the organism, with two exceptions. We have already noted that scientific research on first-order constructs about race as a biological category in science in the last four decades has revealed over and over again that there is greater genetic heterogeneity within versus between major racial groupings (Polednak 1989; Bittles and Roberts 1992; Chapman, 1993; Shipman 1994). One exceptions is that the gene frequencies, as demonstrated in the use of specific polymorphic markers, occur more frequently in certain populations than in others. But this distribution of gene frequencies, though occasionally overlapping with racial groupings, is definitively not only a racially defined issue. For example northern Europeans have greater concentrations of cystic fibrosis than southern Europeans, and both are categorized as "Caucasians." Moreover, southern Europeans have higher rates of beta-thalassemia than northern Europeans—but even more to the point, sickle-cell anemia is found in greater concentration in Orchomenos, Greece, than among African Americans (Duster 1990). This is not a biologically racially defined matter (i.e., racial in the sense of first-order constructs).

Race and "Second-Order Constructs"

Financially, the biggest difference between whites and African Americans today is their median net worth, which is overwhelmingly attributable to the value of equity in housing stock. In 1991, the median net worth of white households ($43,279) was more than 10 times that of the median net worth of African-American households ($4,169, Bureau of the Census: 1991). This is a truth that can be determined by the systematic collection of empirical data, and either replicated or refuted—which is to say that it can be investigated scientifically, without reference to blood groups, the relationship between genotype and phenotype, or the likelihood that one group is more likely to be at risk for cystic fibrosis while the other is more likely to be at risk for sickle-cell anemia. Here is why:

In 1939, the Federal Housing Authority's Underwriting Manual that provided the guides for granting housing loans explicitly used race as one of the most important criteria. The manual stated that loans should not be given to any family that might "disrupt the racial integrity " of a neighborhood. Indeed, the direct quote from Section 937 of the FHA manual went so far as to say that "If a neighborhood is to retain stability, it is necessary that properties shall be continued to be occupied by the same social and racial classes." (Massey and Denton 1993: 54). On this basis, for the next thirty years, whites were able to get housing loans at 3 to 5 percent, while blacks were routinely denied such loans. For example, of 350,000 new homes built in Northern California between 1946 and 1960 with FHA support, fewer than 100 went to blacks. That same pattern holds for the whole state, and for the nation as well.

When people are *systematically* treated differently based upon a social category, it is plausible that they might develop different rates of hypertension than other groups. There is thus the high potential for a feedback loop between social behavior and a biological condition manifested as higher rates of heart disease. It is often necessary to walk a delicate tightrope between describing the feedback between social behavior and biological condition, without leaping to the conclusion that the origins of those differences lie in biological differences. However, to throw out the concept of race is to take the non-thinking alternative—the ostrich approach to race and ethnicity, pioneered and celebrated by the French government: "We don't collect data on that topic. Therefore, it does not exist!"[10] Or perhaps the legacy of Sapir and Whorf is alive and well in the scientific study of race.

Notes

1. For the discussion in this paragraph, and for the references to the German literature that are used here. I am indebted to William H. Schneider (1996).
2. The Human Genome Diversity Project is not to be confused with the Human Genome Project. The latter is a $3-billion effort, jointly funded in the United States by the National Institutes of Health and the Department of Energy. The goal is to map and sequence the entire human genome, and the major rationale for the project, from the outset approximately a decade ago, was to provide information that would assist medical genetics in de-coding, better understanding, and eventually, hopefully producing gene therapeutic interventions

for genetic disorders. In contrast, the Human Genome Diversity Project has been concerned with tracing human populations through an evolutionary history of many centuries. Its goal was primarily to better understand human evolution (Committee on Human Genome Diversity 1997).

3. This was a popular book by Richard Herrnstein and Charles Murray (1994).
4. Tuskegee, after all, was the site of the infamous syphilis experiments on black males—where the Public Health Service of the U.S. Government had studied the racial effects of how the disease ravages the body of blacks in contrast to whites (Jones 1981).
5. Botstein's assertion that the German geneticists "knew" that Nazi claims about Aryan racial purity were false is highly contestable, and the recent work by Proctor (1989) has some abundant counter-evidence.
6. This is a major point that is being made by the two sets of statements about race from UNESCO and the American Anthropological Association, and it can not be repeated too often.
7. Note that the incidence rates of cystic fibrosis estimates are different for Caucasians in Tables 1 and 3. This is in part a function of the fact that there is no general population screen, and estimates are continually re-formulated based upon clinical data, diagnostic skills at different locations and at different periods, etc. Moreover, collapsing the category of "Caucasian" ignores the fact that those of northern European ancestry are at much greater risk for cystic fibrosis than those of southern European ancestry.
8. Whenever there is low incidence, low carrier rates, and low sensitivity of the test—there will be a substantial number of false positives.
9. This was the year that Linnacus published *System Naturae*, in which he revealed a four-part classification scheme of the human races that has residues still today.
10. Perhaps an internally consistent emanation from a society that gave the world the Cartesian formulation about thought and existence—and subject/object dualities.

References

American Anthropological Association (1998) "American Anthropological Association statement on "race," *American Anthropologist* 100(3): 712–13.

Bittles, A.H., and D.F. Roberts, eds. (1992) *Minority Populations: Genetics, Demography and Health*. London: Macmillan.

Burhansstipanov, L., S. Giarratano, K. Koser, and J. Mongoven (1987) *Prevention of Genetic and Birth Disorders*. Sacramento: California State Department of Education. 6–7.

Chapman, Malcolm, ed. (1993) *Social and Biological Aspects of Ethnicity*. New York: Oxford University Press.

Committee on Human Genome Diversity (1997) "Scientific and Medical Value of Research on Human Genetic Variation," in *Evaluating Human Genetic Diversity*. National Research Council. Washington, D.C.: National Academy Press. 16–22.

Devlin, B., and Neil Risch (1992a) "Ethnic differentiation at VNTR loci, with specific reference to forensic applications," *American Journal of Human Genetics* 51: 534–48.

—— (1992b) "A note on the Hardy-Weinberg equilibrium of VNTR data by using the Federal Bureau of Investigation's fixed-bin method," *American Journal of Human Genetics* 51: 549–53.

Duster, Troy (1990) *Backdoor to Eugenics*. New York: Routledge.

Evans, William E., and Mary V. Relling (1999) "Pharmacogenomics: Translating functional genomics into rational therapeutics," *Science* 286(October): 487–91.

Evett, Ian W. (1993) "Criminalistics: The future of expertise," *Journal of the Forensic Science Society* 33(3): 173–8.

Evett, Ian W., I. S. Buckleton, A. Raymond, and H. Roberts (1993) "The evidential value of DNA profiles." *Journal of the Forensic Science Society* 33(4): 243–4.

Evett, Ian W., P. D. Gill, J. K. Scranage, and B. S. Wier (1996) "Establishing the robustness of short-tandem-repeat statistics for forensic application." *American Journal of Human Genetics* 58: 398–407.

Fujimura, Joan H. (1996) *Crafting Science: A sociohistory of the quest for the genetics of cancer*. Cambridge: Harvard University Press.

Gundel, Max (1926) "Einige Beobachtungen bei der rassenbiologischen Durchforschung Schleswig-Holsteins," *Klinische Wochenschrift* 5: 1186.

Hamadeh, Hisham, and Cynthia A. Afshari (2000) "Gene chips and functional genomics," *American Scientist* 88(6): 508–15.

Hermstein, Richard J., and Charles Murray (1994) *The Bell Curve: Intelligence and class structure in American life*. New York: The Free Press.

Hirschfeld, L., and H. Hirschfeld (1919) "Serological difference between the blood of different races," *Lancet* 2(October 18): 675–9.

Jones, James H. (1981) *Bad Blood: The Tuskegee syphilis experiment: A tragedy of race and medicine*. New York: The Free Press.

Katz, S. H. (1995) "Is race a legitimate concern for science?" The AAPA Revised Statement on Race: A Brief Analysis and Commentary, University of Pennsylvania, February.

Kaufmann, Felix ([1944] 1958) *Methodology of the Social Sciences*. New York: Humanities Press.

Klag, Michael, P.K. Whelton. J. Coresh, C. E. Grim and L.H. Kuller (1991) "The association of skin color with blood pressure in U.S. blacks with low socioeconomic status." *Journal of the American Medical Association* 265(5): 599–602.

Marks, Jonathan (1995) *Human biodiversity: Genes, Race, and History*. New York: Aldine de Gruyter.

Massey, Douglas S., and Nancy A. Denton (1993) *American apartheid: segregation and the making of the underclass*. Cambridge: Harvard University Press.

Oliver, Melvin L., and Thomas M. Shapiro (1995) *Black Wealth/White Wealth: A new perspective on racial inequality*. New York: Routledge.

Polednak, Anthony P. (1989) *Racial and Ethnic Differences in Disease*. New York: Oxford University Press.

Press, Nancy, Wylie Burke, and Sharon Durfy (1998) "How are Jewish women different from all other women?" *The Journal of Law-Medicine*: 1–17.

Proctor, Robert N. (1989) *Racial Hygiene: Medicine under the Nazis*. Cambridge: Harvard University Press.

Ries, L.A. G., M.P. Eisner, C. L. Kosary, B.F. Hankcy, B.A. Miller, L. Clegg, B.K. Edwards, eds. (2002) SEER Cancer Statistics Review, 1973–1999, National Cancer Institute. Bethesda, MD. Table XXII-1.

Richards, C. Sue, P. A. Ward, B. R. Roa, L. C. Friedman, A. Boyd, G. Knenzli, J. Dunn, and S. Plon (1977) "Screening for 185delAG in the Ashkenazim," *American Journal of Human Genetics* 60: 1085–98.

Schill, Michael H., and Susan M. Wachter (1993) "A tale of two cities: Racial and ethnic geographical disparities in home mortgage lending in Boston and Philadelphia," *Journal of Housing Research* 4(2): 245–77.

Schneider, William H. (1996) "The history of research on blood group genetics: Initial discovery and diffusion," *History and Philosophy of the Life Sciences* 18(3): 277–303.

Schulman, K.A., J.A. Berlin, W. Harless, J.F. Kenner, S. Sistrunk, B.J. Gersh, R. Dube, C. K. Taleghani, J. E. Burke, S. Williams, J. M. Eisenberg, and J. J. Escarce (1999) "The effect of race and sex on physicians' recommendations for cardiac catheterization," *New England Journal of Medicine* 34(8): 618–26.

Schusterov, G.A. (1927) "Isohaemoagglutinierenden Eigenschaften des menschlichen Blutes nach den Ergebnissen einer Untersuchung an Straflingen des Reformatoriurns (Arbeitshauses) zu Omsk," *Moskovskii Meditsinksii Jurnal* 1: 1–6.

Schutz, Alfred (1973) "Common Sense and Scientific Interpretation of Human Action," in *Collected Papers I: The Problem of Social Reality*, Maurice Natanson (ed.). The Hague: Martinus Nijhoff.

Shipman, Pat (1994) *The Evolution of Racism: Human differences and the use and abuse of science*. New York: Simon and Schuster.

Smith, Edward, and Walter Sapp, eds. (1997) *Plain Talk About the Human Genome Project*. Tuskegee, Ala.: Tuskegee University.

Thompson, E. A., and J.V. Neal (1997) "Allelic disequilibrium and allele frequency distribution as a Function of Social and demographic history," *American Journal of Human Genetics* 60: 197–204.

Vichinsky E.P., A. Earles, R.A. Johnson, M. S. Hoag, A. Williams, and B. Lubin (1990) "Alloimmunization in sickle cell anemia and transfusion of racially unmatched blood," *New England Journal of Medicine* 322(23, 7 June): 1617–21.

Vines, Gail (1995) "Genes in black and white," *New Scientist* 147(8 July): 34–7.

Zerjal, T., B. Dashnyam, A. Pandya, M. Kayser, L. Roewer, F.R. Santos, W. Schiefenhovel, N. Fretwell, M.A. Jobling, S. Harihara, K. Shimizu, D. Semjidmaa, A. Sajantila, P. Salo, M.H. Crawford, E.K. Ginter, O. V. Evgrafov, and C. Tyler-Smith (1997) "Genetic relationships of Asians and Northern Europeans revealed by y-chromosomal DNA analysis," *American Journal of Human Genetics* 60: 1174–83.

8

BSE and the Phillips Report: A Cautionary Tale about the Uptake of 'Risk'*

Kevin Edson Jones

Over the last decade the term 'risk' has become increasingly popular, not only amongst academics in the social sciences, but also in the wider British political arena. The uptake of the language of risk is particularly relevant to the British context in which governments and the public have had to face repeated crises in the regulation of technological, scientific and industrial innovation. The controversies surrounding MMR (measles, mumps, rubella) inoculations,[1] nuclear power, and environmental pollution are all examples of risk issues in the United Kingdom. However, it is agriculture that has provided the context for the most concentrated discussions of risk. 'Mad cow' disease (bovine spongiform encephalopathy or BSE), foot and mouth disease, and the controversies dogging agricultural biotechnology have each staked out a challenge to society about how we contend with risk and scientific uncertainty.

The British government, at the forefront of the need to contend with risk and uncertainty in agriculture, finds itself facing a dual challenge. Firstly, in a climate of increasing industrialisation and technological innovation, the government is expected to make sound regulatory decisions within a context in which scientific advice is routinely inconclusive and the potential for hazard ever present. Secondly, there is an increasing awareness that government cannot rely on science alone in making these decisions, but must furthermore come to terms with

---

*This paper has been developed from a short review of the Phillips report published in the *Canadian Journal Sociology* (Jones 2001).

the social context in which technological controversies take place. Trust in government, science, and industry, changing definitions of expertise and the weighing of the potential benefits of new agricultural technologies with their potential hazards are all key elements of the context in which risk controversies are enacted. Risk, understood in relation to this dual challenge, is revealed not simply as a technical and regulatory problem, but more broadly as a challenge to governance and institutional credibility. The politics of risk, in other words, are also the politics of knowledge, consent and of contingency above all else.

The two-decade-long history of BSE and the government's response to the disease provides one case study of how British social institutions have been able to cope with risk and uncertainty. In particular, this essay will focus on Lord Phillips' report (Phillips et al. 2000a) from the BSE Inquiry, submitted to the House of Commons in October 2000. The task of the Inquiry was to review the emergence and identification of BSE and vCJD in the United Kingdom, along with the adequacy of the actions taken by the government in response to these diseases. Focusing on this case study is, in one part, an attempt to present an account of the strengths and weaknesses of the Phillips report. However, more importantly, this essay seeks to begin a critical dialogue on the way in which governments adopt the language of risk and uncertainty. This involves re-evaluating not only how these terms are applied, but also how we treat them theoretically. By contrasting theories of the risk society with more contextual, or cultural, accounts of risk, a clear need to move beyond technical, scientific, and managerial approaches to risk is exemplified within the limitations of the Phillips report. Governments must move towards an understanding of risk that can be linked to broader cultural and political processes if they are to learn from the BSE crises.

BSE: An Overview

The history of BSE, or "mad-cow" disease as it became popularly known, is simultaneously a scientific and a social story. Neither story encapsulates the history of BSE on its own. Instead, it is in the interplay between the science of BSE and the social response to BSE that we find the heart of the controversy and that which is of the greatest consequence for how we move forward in contending with risk issues in agriculture in the future. This section provides a very brief overview of the primary elements of this story.

The origins of the scientific history of BSE go back to 1985[2] when veterinarians from the then Ministry of Agriculture, Fisheries and

Foods (MAFF) investigated the unusual death of a cow on a farm in Kent. Scientists later diagnosed that the cow had died as the result of contracting a type of disease known as a transmissible spongiform encephalopathy (TSE). TSEs are neuro-degenerative diseases that cause the rapid degeneration of brain cells in the victim. In cattle, animals afflicted with the disease display symptoms of extreme nervousness, hypersensitivity to touch, and loss of balance. Television news footage of cows stumbling, falling over, and being unable to get to their feet became one of the iconic images of the disease (Ford 1996: 18). "Mad cows" were not seen as angry, as connoted by the North American usage of the term 'mad,' but as mentally deranged, insane, and as we will see shortly, as dangerous (Leach 1998: 126–7). BSE is a disease that progresses quickly, induces considerable suffering on the animal, and is invariably fatal (Ridley and Baker 1998).

At the time, as remains the case today, scientific knowledge about TSEs was fragmented and contested. Uncertainty about these types of diseases was compounded by the fact that the veterinary community had not previously encountered TSEs in cattle. Veterinary science did have some experience with TSEs in agriculture as scrapie—the ovine form of the disease—has long been present in U.K. sheep populations. However, despite this familiarity, science, industry, and government were essentially facing a novel veterinary disease of which they had little certain scientific knowledge. Although some scientists speculated that there was a direct link between scrapie and BSE (later proven incorrect), it was unknown why the disease had suddenly developed in cattle and there existed little accurate knowledge to explain how the disease could be transmitted amongst the cattle population.

If the science of BSE was uncertain, the potential hazards the disease posed for human health were even more poorly understood. Scientists in the 1980s had no idea of whether or not BSE would be pathogenic to humans. Veterinary science's experience with scrapie appeared to suggest that the risks posed to humans would be minimal. In the 250 years we have known about scrapie there is no evidence to suggest the disease was ever transmitted to humans by means of the food chain or directly linked to a human form of TSE (Narang 1997: 3). However, scientists did know that it was at least *possible* for TSEs to jump species barriers, possibly from sheep to cows, or from cows to humans. Scientists further postulated that even if scrapie had jumped into the cattle population, there was still no way of knowing what properties the disease would develop in cattle, or whether the disease would even exhibit attributes similar to those of scrapie (Millstone and Zwanenberg 2001: 102). It is upon these shaky foundations that early appraisals of the minimal risk of BSE to humans were generated, appraisals that we now know were ill-conceived. Much of the BSE story therefore pertains to the question of why the government proceeded to act on these misguided assumptions for almost the entirety of the disease's history (Phillips et al. 2000c: para. 11).[3]

In the mid 1990s it became clear that BSE indeed posed a health hazard to the British consumer. A full decade after identifying the disease in cattle, the government announced in March 1996 that scientists had diagnosed a new variant of Creutzfeldt-Jacobs Disease (CJD) in ten young people and that the disease was strongly linked to BSE. Despite the rarity of CJD, medicine had been actively treating those with the disease since it was first identified in the 1920s. However, the new variant of the disease (vCJD) was markedly different from CJD. Those inflicted with vCJD are significantly younger, display psychiatric symptoms much earlier and suffer a longer duration of illness than those suffering from traditional CJD (Phillips et al. 2000d: paragraph 6.1). Pathologically, those suffering from vCJD display symptoms very similar to cattle suffering from BSE. The first signs of the disease are seen in changes in behaviour and mood (e.g., depression). Often, at the early stage of the disease, victims with these symptoms were diagnosed incorrectly and referred to psychiatric treatment (Phillips et al. 2000d: para. 6). As the disease progresses victims suffer from debilitations in movement and from loss of memory (Narang 1997: 222). vCJD, like BSE, is fatal without exception.

The 1996 announcement brought public anger over BSE to a head. Media coverage, which had lulled in the early 1990s, exploded and BSE and vCJD again dominated the newspapers and TV news programming (Kitzinger and Reilly 1997: 343–4). Many in the public, with their fears over the safety of British beef now confirmed, stopped buying beef or went further and joined the swelling ranks of British vegetarians. Industry, already suffering from this sudden loss of domestic confidence in British beef, suffered a further blow on March 1997 when the Commission of the European Union banned the export of all British bovine products (European Commission 1996). Facing an angry citizenry, the potential downfall of the British livestock industry and European sanctions, the government came under increasing pressure to halt the disease and restore faith in British industry and in government institutions, policy and regulation.

Phillips and Risk

Almost two years after the government's announcement linking vCJD and BSE, and within a context of public anger and mistrust. Agriculture Minister Jack Cunningham announced the establishment of the BSE Inquiry to parliament in December 1997. Chaired by Lord Justice Phillips, the BSE Inquiry was given the following mandate:

> To establish and review the history of the emergence and identification of BSE and variant CJD in the United Kingdom and of the action taken in response to it up to 20 March 1996; to reach conclusions on the adequacy of that response, taking into account the state of knowledge at the time; and to report on these matters to the Minister of Agriculture, Fisheries and Food, the Secretary of State for Health and the Secretaries of State for Scotland, Wales and Northern Ireland. (Phillips et al. 2000b: xvii)

The Inquiry was asked for a step by step account of why it took government and industry so long to gain control over the disease in cattle and acknowledge the potential hazards it posed to humans. However, in producing a response to this rather straightforward, although difficult, task the Inquiry ran up against some persistent and much less straightforward questions about the ability of social institutions, and governments in particular, to contend with the complexity of modern society. This included an interrogation of the ability of government and industry to react to the potential hazards proposed by complex technical and industrial systems of production that are increasingly characteristic of modern agriculture in Britain. Phillips recognizes that this matter is further complicated by the fact that we are seldom fully, if at all, aware of the dangers we face from these systems. When the mandate refers to the "state of knowledge [about BSE and vCJD] at the time," it refers not only to what scientists knew and did not know about BSE and vCJD. The mandate also might be read as the need to take account of the government response to BSE according to what science could not be expected to know or may not even be able to know; the "unknown unknowns" as Grove-White (2001) refers to them.

Approaching the Tenets of the "Risk Society"

In addressing this mandate and conveying the outcomes of the BSE Inquiry to the public, the Phillips report relies heavily on concepts of risk and uncertainty. In doing so, the report most closely reflects notions of risk associated with the work of Ulrich Beck (1992) and Anthony

Giddens (1990). Risk, according to these authors, is a catch-all term that characterizes the rapid and profound changes society is undergoing as it is transformed from one stage of modernity to the next. Where some authors speak about risk and vulnerability as the inevitable products of increasingly complex technologies (e.g., nuclear power; Perrow 1984; Murphy 2001), Beck and Giddens extend this to speak of risk as the inevitable outcome of increasingly complex forms of social organization (e.g., globalization). Hence, the authors propose that we are now living in a "risk society." A reality that both authors suggest has been epitomized by the BSE story (Beck 1998; Giddens 1998).

I do not wish to argue that Phillips has full-heartedly embraced ideas of the risk society in reporting the outcomes of the BSE Inquiry. However, terms such as 'risk' and 'uncertainty' become overarching themes throughout the report, and although Beck and Giddens in no way hold a monopoly on the use of the terms, Phillips does appear to tentatively approach several of the key tenets of their theories.[4] Foremost amongst these is the argument that society must recognize and seek to come to terms with the fallibility and inconclusiveness of science. Knowledge, as many have already noted in the sociology of science, is always uncertain (e.g., Latour 1987). As such, the argument runs that we cannot faithfully accept that scientists have all the right answers, or are in a position to protect us from all of the potential dangers we face in our everyday lives. In the risk society, science and scientific authority have started to lose their sanctity (Giddens 1994: 87–8). Secondly, proponents of the risk society argue that the hazards facing society today are the unpredictable consequences of human actions and not the result of some external natural power. Plainly stated, society's attempts to dominate nature have created novel economic risks, global social risks, and health risks with which society must now contend (Beck 1992: 80–4). Finally, within this context of risk and epistemological uncertainty, traditional social institutions are seen as no longer able to cope with the protective tasks with which they have been assigned and on which society depends (Beck 1996). The risk society demands new forms of governing the increasing complexities (scientific, industrial, economic) of today's world.

The Cause and Spread of BSE in Cattle

To recall, when BSE was first identified in 1986 the veterinary sciences were faced with a novel disease in cattle and knowledge about the nature of the disease was sketchy at best. Although familiar with the family of diseases of which BSE is a part (TSEs), certainty about the cause of the disease and why it suddenly appeared in cattle continues to elude science even today. Early in the BSE story there were several theories that sought to explain where the disease had come from. Some suspected BSE entered Britain through the importation of cattle, others pointed the finger at the use of veterinary vaccines and pharmaceuticals, and yet others blamed some sort of BSE virus (Ford 1996: 153–73). Early on, the theory that gamed the greatest acceptance assumed that the source of BSE was ovine scrapie, and that the disease had somehow jumped the species barrier between sheep and cattle. Yet, all these theories are now believed to be spurious and the Phillips report instead categorizes the cause of BSE as "unknown," and acknowledges that TSEs may develop sporadically (randomly and without explanation) in species in which they have not been identified previously (Phillips et al. 2000b: 249–50).[5]

However, if the cause of BSE is recognized as uncertain, or potentially even unknowable, the report is far more certain in the ascription of a cause to the persistent and widespread transmission of the disease throughout the U.K. beef industry. As early as 1988, this epidemic side of the BSE story was attributed to the use of high-protein animal feeds containing meat and bone meal derived from infected cattle. Intensive agriculture, it turns out, had turned cows into carnivores and cannibals. This practice, unregulated for several decades, constituted what Phillips refers to as "a recipe for disaster" (Phillips et al. 2000b: xvii).

So, if the pathway of infection was known early in the BSE story, why was the disease not properly contained for a further decade? For Phillips the answer is clear. The government made mistakes in ascertaining the extent of the problem and in expressing the resolve necessary to deal with it. These mistakes are attributable to the government's tendency to rely on putative knowledge about the transmission of the disease when 'facts,' as such, were unavailable. In other words, politicians and government regulators continued to seek to operate on the basis of certainty within a context of uncertainty. Yes, they knew the pathway through which the disease was being transmitted, but the actions taken to stop the disease were based upon several false assumptions about the pathway's operation.

In particular, Phillips is critical of the government's over-reliance on the implementation of a ban on the use of recycled ruminant proteins in ruminant feed, put in place in 1988. The logic behind the ban seemed straightforward. If BSE was being transmitted through the use of protein feeds, and possibly even linked to feeds consisting of material from scrapie-infected sheep, then the solution lay in cutting these feeds out of the food chain. However, we now know that the original feed ban was a spectacular failure. At the heart of the problem was an assumption that a relatively large amount of infected material needed to be consumed by a cow in order to contract the infection. Consequently, the ban failed to address the perpetuation of BSE that was occurring in feed lots as a result of the cross-contamination. Pig and poultry feeds were not addressed in the ban and were still being produced with material derived from sheep and cattle. Small quantities of these feeds were inadvertently being mixed with cattle feed that, under the ban, was supposed to be free of all ruminant proteins. At the time of the ban, because of the belief that the disease could not be transmitted through the consumption of small amounts of infected material, cross-contamination was perceived as constituting a minuscule risk and therefore discounted. What in effect the government was doing was employing regulatory measures on the basis of unproven beliefs about the nature of the disease without acknowledging the uncertainty of these assumptions (Phillips et al. 2000b: xxi–xxii, 18, 255).

It should be noted that, although Lord Phillips is decidedly uncritical in his investigation of the role of industry influence in the government's production of regulatory measures, the decision to not extend the ruminant feed ban to pig and poultry feeds did not take place in a political vacuum. Instead, as Millstone and Zwanenberg (2001) note, significant pressure was being felt by the government to avoid making regulatory decisions that would damage industry concerns. Nobody knew the consequences of feeding pigs and chickens infected bovine material and nobody foresaw the extent of the problems being created by cross-contamination. However, when advice was put forward that did acknowledge the uncertainties presented by the disease, and that favoured precautionary action over inaction, the government chose to ignore or suppress this advice. When the Southwood Working Party, established to advise the government in the early stages of the BSE story, seemed on the verge of calling for a total ban on ruminant feeds, they suddenly lost their nerve. The recommendation to extend the ban was dropped in favor of what has been cited as an attempt to preserve the rendering industry's principal market—chicken and pig feed (ibid.: 105).[6] Uncertainty not only led the government to ignore the unknown risks posed by BSE, but also allowed politicians and regulators to avoid making decisions that would have gone against the economic interests of the animal processing industry.

When, in late September 1990, responses to cross-contamination were put in place (e.g., a ban on the use of specific bovine offal, or SBO, in all animal feeds), the transmission of BSE continued on the back of further misguided assumptions about the disease. Materials that were wrongly assumed to not be infectious, such as eyes and lymph nodes, remained outside the ban. In the abattoirs, the supposition that a large amount of material was needed to transmit the disease persisted, resulting in a lackadaisical attitude toward the identification and separation of SBO during the slaughter of an animal. Furthermore, the government lacked a clear mandate and structure to inspect slaughterhouses and trusted industry to self-police regulatory directives.[7] More galling still, with no external enforcement of the SBO ban, industry intentionally passed contaminated material off as "clean" in several instances (Phillips et al. 2000f: 4b). In general, Phillips states that these problems stemmed from a prevailing lack of appreciation of the potential risks created by BSE and of the importance of the measures being prescribed. It was not until almost a decade after the source of the BSE epidemic had been identified that the government finally introduced further measures to shore up these problems and contain the disease.

However, if the cause and spread of BSE were characterized by scientific uncertainty and the inability of government to take appropriate regulatory action within this context, what of the circumstances leading to the appearance of the disease itself? The Phillips report is very direct in answering this question, attributing the source of the disease to human innovation and industrial processes. In this light, Phillips identifies intensive farming practices and industrial models of production as the causes of the BSE epidemic (Phillips et al. 2001b: xvii). Foremost amongst these practices was the necessity of using recycled animal proteins in livestock feed to nourish higher-producing animals, with very little consideration of the consequences of these actions. Uncertainty, in this sense, is not understood solely as a limitation of knowledge, but as the source of potential hazards, or risks, that are the yet unknown products of increased complexity and technological development that characterize modern agriculture, in making this conclusion the Phillips appears to agree with Beck and Giddens' descriptions of the risk society and late modernity: "In a primitive society, the major hazards are those posed by nature. In a complex modern society the acts of individuals or corporate bodies may also involve serious hazards to other members of society" (Phillips et al. 2000b: 31).

In this statement, Phillips makes a move to tie the scientific uncertainty surrounding the disease and the uncertain actions of the government with some broader notion of what we might consider the 'risk society.'

The Transmission of BSE to Humans

The government's ability to contend with uncertainty and risk also dominates the Phillips report's treatment of the link between BSE and vCJD. Like the above discussion of the handling of the BSE epidemic, the Phillips report is critical of the government's lack of a serious engagement with the human health risks posed by BSE and the tendency to rely on false certainties in guiding its response. Here the role of the Southwood Working Party and the government's handling of expert advice comprise one of the key themes of the human health side of the BSE story.

Jointly responsible to MAFF and the Department of Health (DH), the Southwood Working Party was established in May 1988. Chaired by Richard Southwood, professor of zoology at Oxford, the committee was composed of a collection of scientists who were both academically well-regarded and had considerable experience working in advisory capacities. The committee was given the broad task of examining the implications of BSE for both animal and human health (Southwood 1989) and was expected to provide both scientific and policy advice to the government (Millstone and Zwanenberg 2001: 104). Over the course of the BSE story the committee produced a series of interim recommendations. These advisements, which were subsequently implemented by government, included the compulsory slaughter and destruction of all cattle showing infection and a ban on the use of "specified offal" (e.g., brain and spinal cord) in human food. The committee produced a controversial final report in February 1989 that concluded that in its estimation, although the transmission of BSE to humans was possible, it was only remotely so (Phillips et al. 2000b: 50–55).

The majority of the criticisms made against the Southwood Working Party concern the government's handling of the committee's advice, rather than the work of the committee itself. In particular, Phillips lambasts the government's blind acceptance of Southwood's conclusion that BSE posed only a minimal risk to human health and the government's tendency to cite this conclusion as constituting a scientific risk appraisal (ibid.). The government forged a rigid position that argued that no further precautions were needed. As the Phillips report accounts: "the conclusions of the Southwood Working Party were not reviewed. Their recommendations were treated not as advice, but as definitive of the precautionary measures which did, and did not require to be taken" (ibid.: para. 1221). Moreover, the government repeatedly used the South wood findings to discredit and deride the opinions of dissident scientists, thus polarizing debates over the human health risks of BSE and further hardening the government's position (ibid.: para. 1182; Phillips et al. 2000g: sec. 5). Furthermore, the government outwardly held up the objective truthfulness of scientific certainty, while simultaneously exerting pressure to determine what these certainties would be. The South wood Working Party was cast as the voice of scientific certainty, but was repeatedly advised to be cautious in its advice and wary of the consequences its decisions would have on the beef market. In summary, what was considered scientific and what was considered risky had become the stakes of a highly politicized engagement between the government, industry, and the expert working party (Millstone and Zwanenberg 2001) This approach to the committee's advice continued long after serious doubts had compromised many of the assumptions about the hazards posed by the disease (Phillips et al. 2000b: paras. 1209, 1221).

Foremost amongst these assumptions was the perceived relationship between scrapie, BSE and CJD. Simply, it was believed that because scrapie had such a long history in the United Kingdom and could not be directly linked to CJD in humans, BSE would act in a similar manner and therefore was not likely to pose any serious hazards for human health. In reply to a letter from a neuropathologist who had raised concerns about the South wood report's handling of the human health risks posed by BSE, Sir Richard Southwood put it this way:

> As you can imagine, in this report it was extremely difficult to steer the proper course between causing excessive alarm and undue complacency. The evidence to date seems to indicate that the BSE agent is very similar to scrapie, and of course we have lived with scrapie for two hundred years, and most of us have at some time or other eaten sheep offal—though the incidence of CJD remains low. It was this line of argument that finally convinced us not to press the point that you have made in your letter any more strongly. (Phillips et al. 2000c: 56)

A sense of apathy grew out of Sir Richard's speculations about the link between BSE and scrapie, and like the circumstances surrounding the spread of BSE amongst cattle, little was done to investigate the possible sources of risk to human health. Once again it was assumed that, even if the disease was theoretically able to jump species, a relatively large amount of infected material would need to be consumed to do so. Likewise, legislation banning the use of SBO in human foods was considered more than sufficient. Little concern was given to whether industry would effectively police the ban, or whether these materials could physically be separated without contaminating human foods in the first place. We now know that scrapie and BSE are not directly linked and that the regulations put in place to prevent contaminated material from getting into the human food chain were fully inadequate.

The consequences of these failures were dramatic. Between 1989 and 1996, the British public was continually exposed to a deadly disease, due directly to government and industry mishandling of knowledge and uncertainty. To date, more than 100 Britons have died of vCJD. As the incubation periods of the disease remains unknown, estimates of how many more will be afflicted with the disease are inconsistent and have produced numbers ranging from a few thousand to hundreds of thousands (Sandra Blakeslee, "Estimates of death toll from mad cow disease vary widely," *New York Times* 30 October 2001; Phillips et al. 2000b: 258–9). On top of this, the agricultural industry has been tainted by the BSE scandal, and more recently by foot and mouth disease. The livelihoods of farmers, the sustainability of the agricultural industry, and the welfare of the British countryside have all been tainted by a disease that was allowed to progress unchecked (Woods 1998).

By the time the government finally overcame their spurious assumptions about the risks posed by the disease and recommitted itself to curbing the situation, things had gotten out of control. The widespread infection of the cattle population and intense public anger over mishandling of the disease had progressed to such an extent that those in office felt it necessary to take dramatic steps to cleanse 'mad cow' disease from Britain. Between March 1996 and the end of 1999. Phillips reports that more than 3.3 million cattle were prematurely slaughtered in the United Kingdom (Phillips et al. 2000b: 21).

In conclusion, Phillips is decisive in stating that the human health tragedy of BSE was not directly attributable to the uncertainty of the science of BSE and vCJD, but to the government's mishandling of this uncertainty. By relying on a series of false assumptions it produced an unstable and dangerous platform on which it made, or failed to make, the regulatory decisions necessary to avoid disaster. Gavin Little offers us this concise summary of the situation:

> The end result . . . was a collective failure over the period 1989–1996 to give enough weight to the possibility that, given the lack of knowledge about BSE, scientific proof of its ability to "jump species" lo humans might only emerge once consumers were infected, which is too late for effective regulation. Against a backdrop of political and commercial pressures, there was . . . insufficient appreciation on the part of government and scientific advisers of the complex . . . nature of the risks posed by BSE, or of the "built in ignorance of science towards its own limiting commitments and assumptions." Regulatory action was therefore taken on the basis of a deferential, non-precautionary reliance on what was erroneously taken to be an authoritative, objective and definitive scientific risk assessment. (Little 2001: 747)

Trust as a Casualty of BSE

The government's failure to create effective regulatory decisions within a context of risk and uncertainty is also mirrored in its relationship with the British public. Trust in government, science, and industry were all severely challenged by BSE and resulted in a breakdown in relations between these institutions and the broader public. As the Phillips report put it, "trust" became another casualty of the BSE story (Phillips et al. 2000b: xviii).

Although without wishing to fault individuals, Lord Phillips is highly critical of what he describes as the undertaking of a "campaign of reassurance" in the communication of risk to the public (Phillips et al. 2000a: 261). Beef, the British public was routinely assured, was not only safe, but good for you as well. When coupled with a reliance on false scientific certainties about BSE and CJD, such assurances were made with a clear conscience and allowed government officials to cling to a belief in the validity of their decisions and the decision-making process. Such confidence led the government to go to absurd lengths in defending its decisions to the public. One image that stands out as representative of this approach is that of former Agriculture Minister John Gummer feeding his daughter a hamburger on national television as part of a much maligned attempt to quell public "hysteria" over BSE.[8] Gummer explains the rationale behind his actions in the following terms:

> In matters as important as these it is essential to have a personal benchmark to be applied to decisions wherever appropriate. In such

circumstances I applied the test: "Would I be entirely happy for my children to eat this?" That seemed to me to be the proper question for a non-expert to ask when assessing the views of experts. At all times I saw my primary role as protecting the public. (BSE Inquiry 1998: para. 13)

However, Phillips goes further than critiquing the government's tendency to over-rely on false certainties when communicating risk to the public. Moreover, the BSE Inquiry found that communication between the government and the public was guided by a pervasive fear that an anxious and potentially hysterical public would respond irrationally to the risks posed by BSE. Politicians and government officials feared that speaking about potential risks—the unknown unknowns—would lead the public to blow these risks out of proportion and cease to purchase U.K. meat products. The report does not propose that there was an attempt to cover up the hazards posed by BSE in order to protect the beef industry, as might be suggested by more cynical observers of the BSE story. Instead, Lord Phillips documents that scientists, government officials, and industry representatives, when faced with uncertainty about the disease and intense public pressure, routinely resorted to more familiar statements of certainty and reassurance (Phillips et al. 2000b: 232–5). As one witness testified to the Inquiry:

> Given the strength of public debate on the matter at the time one was aware of slightly leaning into the wind. You could not just stand upright and give a totally impartial view of what was the situation. There was a strong danger of being misinterpreted one way or the other, and we tended to make more reassuring sounding statements than might ideally have been said. (ibid.: 265)

In summary, several themes have emerged from Phillips' critique of the government's handling of the spread of BSE amongst cattle, its link to vCJD and human health risks, and the communication of risk and uncertainty to the public: themes I suggest closely reflect several of the tenets set out by the sociology of the 'risk society.' Phillips has told the BSE story as a tale of uncertain knowledge and risks. Risks, the report identifies, although somewhat circumspectly, as the consequences of the complexities of modern agriculture and intensive agricultural systems. Above all, Phillips argues that BSE is the failure of government to work within a context of risk and uncertainty to protect society from the diverse hazards posed by the disease.

"Lessons Learned"—Risk Management and Communicating Uncertainty

The overarching priority, or 'lesson learned,' that comes out of the Phillips report, is the necessity for social institutions to be better prepared to contend with risk and uncertainty. Two points stand out as particularly relevant to this discussion of BSE and permeate the majority of the report's findings. The first makes the identification and management of risk the priority in contending with future uncertainties. The second pertains to the need for governments to be able to openly and effectively communicate uncertainty to the public.

Managing Risk

At the base of the first of these priorities, what I refer to as the management of risk, is the basic insight that 'uncertainty can justify action' (Phillips et al. 2000b: 254). The report is emphatic that social institutions can no longer remain complacent about the potential risks associated with the application of modern science and technology, even though these risks may often appear remote. Instead, science, government, and industry must each ensure that all necessary precautions are taken to both identify and contend with hazards regardless of their assumed probabilities.

Thus, although lauding the individuals who first identified BSE and linked the disease to human health risks, the report strongly asserts that organizational improvements are still needed to better *identify* future sources of risk to British society. On one hand, this means increasing cooperation between government and industry in order to improve the nation's animal disease surveillance systems (ibid.: 251–2). On the other hand, it also means redressing deficiencies in the standards and availability of the scientific expertise needed to expeditiously recognize and track potentially dangerous veterinary diseases (ibid.: 225).

The Phillips report, furthermore, suggests that contending with uncertainty also means ameliorating the way in which government, science, and industry work together to address risks once they have been identified. Specifically, the report asserts that communication and cooperation between these institutions must be improved if effective and timely actions are to be taken in the future. This includes a re-evaluation of relations within government, between government and science, and between government and industry.

Within government specifically, the Phillips report foregrounds the need to promote greater intergovernmental communication and cooperation. This necessity is highlighted by the failure of MAFF and the DH to work together 10 recognize that BSE was not exclusively a veterinary disease, but a threat to human health as well (Phillips et al. 2000b: 235–6). Likewise, the report urges the government to re-evaluate its use of scientific expertise. Risk management, in this context, means better management of science and is to be accomplished by better coordinating and funding scientific research, and ensuring that the government is able to access appropriate and skilled sources of scientific expertise. Particular attention is paid to the government's misuse and over-reliance on expert committees, such as the Southwood Working Party. The report argues that if it is necessary to resort to the advice of external experts, then care must be taken to ensure that not only their conclusions are clearly articulated to government, but their limitations in knowledge and assessments of risk as well. Furthermore, efforts must be made to avoid the temptation to consider the conclusions of expert committees as determinative of policy (ibid.: 238–41). Finally, in response to growing concerns about the relationship between government and industry, Phillips asserts that governments must provide and enforce clear industrial regulations aimed at the prevention and treatment of risk. These regulations must not only address those risks considered to be "reasonably probable," but also address those that are considered only "mere possibilities" (ibid.: 266–72).

Communicating Uncertainty

In response to the conspicuous misrepresentation of risk that characterized the BSE story, the Phillips report makes it clear that providing security from risk also entails improving communication between social institutions and the public. Governments, scientific experts, and industry representatives are all urged to clearly communicate uncertainty and potential risks to the public. This includes the clear conveyance of the incomplete nature of knowledge upon which decisions about the probability of risk are being made, as well as the constraints these gaps imply for the ability of social institutions to secure society against these hazards. As a prerequisite to building this foundation of openness, Phillips asserts that it is necessary to first appreciate that the public will react rationally when provided with an honest appraisal of risk and uncertainty. In effect, the report is impelling government to make their practices more transparent and to ensure that in future the importance of precautionary measures will not be played down on the grounds that the risks they address are unproven. The goal of this openness is to regenerate public confidence in the institutions tasked with providing security and well-being to society.

Limitations of Phillips and the "Risk Society"

On one hand the Phillips report should be congratulated for incorporating an understanding of risk and uncertainty into its account of the BSE story. However, these successes are limited by the failure to acknowledge the broader implications of these terms. As a consequence Phillips socially decontextualizes risk and uncertainty and in doing so upholds the status quo instead of addressing the social conditions at the heart of the BSE story. Perhaps this is not particularly surprising, but by doing so, Phillips is compelled to treat BSE as a technical problem requiring better technical management. Phillips offers little in the way of alternatives to the institutionalized authority of science or the agricultural practices that the report itself so heavily implicates in the BSE story.

Risk, Context and Contingency

To recall. Beck and Giddens argue that risks are the outcome of increased social and technological complexity, stressing their uncertainty, their social origins and the problems they create for the institutions charged with protecting society from these risks. Phillips has drawn on related ideas to both describe the BSE story and to reach some conclusions from which government can hope to learn. However, Phillips does little to engage with the social backdrop that lies behind the ideas of the risk society; in this case, the state of modern agriculture and the status of science in Britain.

One might be tempted to accept these limitations in social scope under the pretense that the BSE inquiry was not the appropriate forum to address such broad social concerns. It could be the case that Phillips may have chosen not to delve further into these broader social cases in order to avoid detracting from his conclusions: the need to push governments to look for new ways to operate within contexts of uncertainty and to force them to do so in a more open and transparent manner.

The failure to engage with the social context behind the *creation* of BSE can also be related to the conception of risk that Phillips adopts and, I have suggested, closely relates to theories of the risk society. Risks, for Phillips, are the *inevitable outcomes* of an increasingly complex social and technological world with which society must now contend. The emphasis, however, is placed on contending with risks and not directly challenging the context behind their creation. Giddens, for example, directs our attention toward the "consequences of modernity" (1990). He proclaims that modernity—a historically specific "mode of social order or organization" (ibid.: 1)—is undergoing a radical transformation that, although offering benefits, creates new risks and dangers for society. The rapidity and scope of these transformations are such that Giddens characterizes modernity as a world that has begun to hurtle out of control: "Living in a modern world is more like being aboard a careening juggernaut rather than being in a carefully controlled and well-driven motor car" (1990: 53).

Accordingly, Giddens is preoccupied with providing a descriptive account of these changes, their consequences, and the urgency for social institutions to respond to the potential downside of the modern transformation. Modernity is something that society must face and grapple with and consequently little scrutiny is given to modernity as the focus of a contestation in and of itself. As David Miller points out the presentation of risks as the "inevitable concomitants of technological and cultural developments" threatens to leave risks "in the grip of political quietism" (Miller 1999: 1239). Within Giddens' modernity a politics of risk is largely responsive and centers on how we contend with technological and cultural change. It does not allow us to imagine political spaces in which the social shaping of technology and culture are contested. Alternatively, if we imagine modernity as a context and not as an absolute, then risks are not inevitable but are contingent. What types of knowledge and technology society chooses to pursue, how we apply and organize them, and the types of risks they create are not determined, but can be understood as the outcome of specific social, political, and economic relations and decisions. The politics of risk thus can be shifted from a singular focus on contending with the consequences of risk to address the circumstances that give rise to risk.

Risks should not be construed as the inevitable outcomes of modernity and its associated uncertainties, but in association with the social and cultural politics that lead up to their creation. To believe that society must accept that risks are inevitable directs us to believe that little can be done to avoid them other than to take the protective measures to prepare for them in order to limit their occurrence. To believe that risks are socially constructed permits us to challenge the circumstances leading to their appearance, to interrogate the type of society we wish to inhabit, and most importantly to seek out alternatives (ibid.: 1250–53).

BSE, Consent, and the Intensification of Agriculture

The biggest challenge facing the BSE Inquiry was coming to terms with the technologies, processes, and organizations that shape modern agriculture in Britain. As Phillips clearly states, it is intensive and industrialized agricultural systems that are to blame for giving rise to BSE: "BSE developed into an epidemic as a consequence of an intensive farming practice—the recycling of animal protein in ruminant feed" (Phillips et al. 2000b: xvii). Yet, despite making such a strong statement, the report does little to challenge these practices themselves. Instead, Lord Phillips is content to focus on how these practices are employed, regulated, and managed.

Phillips' limitation in the scope of his engagement with modern agriculture can in part be related to the BSE Inquiry's uptake of the notions of risk and uncertainty. By taking modernity for granted and accepting risks as inevitable, Phillips is directed toward prioritizing the management of the *consequences* of uncertainty. As a result, he displays little regard to the potential for reducing the a priori social production of risk. The conclusions that the BSE Inquiry has produced almost always are reduced to the need to develop better management and regulatory strategies around technological uncertainties and the potential hazards they pose.

Certainly risk management is an essential part of the governance of new agricultural technologies and practices, however, it also limits the critical potential offered to government by conceptions of risk and uncertainty. To begin with, by placing risk within a management framework. Phillips has enclosed risk within very traditional models of governance. What sets these models apart is their historical tendency to stress scientific and technical appraisals when regulating agricultural practices and determining risks. This accentuation of the techno-scientific almost always is at the cost of developing a serious engagement with the social practices that underscore innovations in agriculture.

As a result, instead of addressing agriculture as a social practice, Phillips obsesses over the minute technological, scientific, and industrial processes behind the risks associated with BSE. One is hard-pressed to find a discussion of agriculture itself, and is instead presented with page after page outlining the particularities involved in the slaughter, rendering and distributive use of an animal. BSE, according to Phillips, is a story that pertains to feed cross-contamination, the failure to classify and separate infected from non-infected materials and the ability to create and enforce regulations to solve these problems. At its most mundane moments, the attention of the Inquiry is reduced to discussions of the different methods of splitting a carcass, dying and removing offal, or de-boning a skull (Phillips et al. 2000h).

Certainly, Phillips should be lauded in arguing that governments need to recognize the routine uncertainty associated with scientific risk assessments and technical processes. However, science and technique are almost exclusively presented as the means with which society is going to be able to protect society from the inevitability of future risks. Contending with risk and uncertainty in agriculture is confined to doing better science and better regulating industrial and technical processes.

I do not wish to suggest that the conceptions of risk and uncertainty Phillips adopts are deterministic of these sorts of techno-scientific conclusions. Rather, by taking modernity for granted and treating risks as inevitable, Phillips is easily able to fit notions of risk into the management frameworks that prioritize scientific knowledge and expert systems of decision making.

A helpful alternative to this conception of risk that directs our understanding of BSE beyond this narrow-sighted focus can be found in the work of Mary Douglas and Aaron Wildavsky. They define risk, as partly a problem of the certainty or uncertainty of knowledge (Douglas and Wildavsky 1982: 5), but more importantly link risks to issues of consent over the "most desired prospects" for the future. Risks, in other words, do not simply reflect the complexities of modern society, but are cultural concepts around which the future of society is being shaped. What we choose to see as risks and what risks we ignore are moral and political choices. Risks are not socio-technical inevitabilities. Where uncertainties of knowledge suggest the need for technical solutions and further scientific research, there are no clear solutions for contending with uncertainties of consent. Instead, issues of consent create spaces of overt political contestation, which connect debates over science and technology to a broader cultural and normative context.

With this perspective BSE can be understood as representing a broader debate over what we consider acceptable forms of agriculture and food production in contemporary Britain. Cultural conceptions of risk would suggest that the cross-contamination occurring in feed lots should not be understood solely in terms of the practical processes by which potentially infected material was being identified and removed from the carcass and the food chain, or in terms of the regulation of these practices. Rather, the contamination issue is representative of a broader concern for the risks posed by the practice of feeding animals to one another in the first place. Stepping back from techno-scientific risks allows us to see the BSE story as part of a contestation over the industrialization of agriculture. Public responses to BSE and vCJD on the one hand pertain to fears over the safety of the food supply and the government's ability to protect society from these risks. On the other hand, the anger that erupted in response to BSE furthermore marks out a battle of consent over the increasing intensification of farming. This is precisely the point where the BSE Inquiry pulls up short in its investigation. Simply stated, if the BSE epidemic was caused by industrialized agriculture and intensive livestock farming, as suggested by the Phillips report, should attention not be granted to the employment of these practices and not just the processes by which they operate? These were certainly the expectations that many in the United Kingdom voiced going into the BSE Inquiry. Emily Green, writing in the *New Statesman*, voices the expectation that with the establishment of the BSE Inquiry, the future of agriculture was hanging in the balance ("In this inquiry's hands: The future of agriculture" 9 January 1998). In hindsight, such expectations were obviously misplaced.

However, if the BSE Inquiry has not been able to interrogate the broader connotations of risk attached to BSE, the British public has been much more capable of doing so. Whilst the Phillips report tells the story of BSE within a bounded vision of risk, members of the general public have been less inclined to narrate the story in this way and were able to extend an awareness of BSE far beyond the implications and dangers of the disease itself. For many, the risks of BSE related to exoteric reflections and overt contests over the condition of contemporary British Society. For example, concerns raised over the condition of the national food supply were not limited to the contamination of beef by "rogue prions," but also included debates over developments in genetic engineering and the commercialization of genetically modified foods. Likewise, media vocalizations of a growing distrust in science pertained to much more than a discussion of the scientific handling of the disease itself, but more broadly evoked concerns about the increasing encroachment of industry into science and government.[9] Similarly, vegetarianism, the ethical treatment of animals, organic farming, and the plight of rural Britain were all issues that the public linked to BSE but do not take up any significant space in the thousands of pages that comprise the Phillips report.

Democratizing Government—Science—Citizen Relations

The limitations created by Phillips' managerial and technocratic focus on risk are also reflected in the report's approach to the communication of risk and uncertainty to the public. Although the report takes a positive step in promoting an open and transparent model of communication, it is limited by the adoption of conservative assumptions about the nature of knowledge and expertise. In particular, Phillips' failings can be traced to the separation of these terms from a clear understanding of the social application of power that characterizes relations between government, science and the public. In doing so, the report not only fails to challenge the way we understand citizen-science relations, but risks upholding the same values that gave rise to the failures of communications endemic to the BSE crisis—the maintenance of science as the lone authoritative voice in the regulation of agriculture.

At the center of the Phillips report's approach to the communication of risk and uncertainty is an assumption that trust can be generated by simply improving communicative processes and developing an awareness of the rationality of the public at large. However, the Phillips report continues to cast science as the *exclusive procurer* of knowledge, while at the same time upholding a perception of the public as *passive receivers.* As a consequence of this monological approach to communication, the public's failure to understand science and the true nature of risk become the predominant concern in the relationship between scientific experts and the public. Thus, although the report recognizes that the public should not be assumed to be irrational, they are not recognized as significant sources of knowledge themselves. Instead, the public are derogated as *tabula rasa*—blank slates (Irwin, Dale, and Smith 1996: 48; Michael 1996: 109).

Consequently, these unidirectional assumptions about the relationship between experts and the public have isolated the majority of British citizens from participating in social responses to risk and uncertainty. Although the report does state that "a lay member can play a valuable role on art expert committee" (Phillips et al. 2000a: 262), the contextual experience and knowledge that the public might bring to the table are for the most part ignored. For example, the initial failure of the government's ban on the inclusion of certain specified offal in animal food products is representative of the wider failure to consider the importance of contextual knowledge. Scientists and government officials knew that, in order to stop the infection of other animals, it was necessary to remove those parts of the animal considered most likely to contain the infected material. Within the laboratory and in bureaucratic offices the ban would seem to be a relatively straightforward matter. However, to place the ban in context requires that we consider it relation to the messy and chaotic character of the abattoirs in which the ban was to be effected (Irwin 1995: 116; Phillips et al. 2000h). No consultations were held with either workers in the abattoirs or with other members of the public who were actively involved in the meat and livestock industry. Surely, these voices could have made an important contribution to generating effective policies that took account of the context in which any ban would be implemented? By failing to reflexively consider the limitations of their expert knowledge and acknowledge the potential contributions of contextual forms of knowledge, the government allowed the transmission of the disease to continue unimpeded.

The general disinterest the Phillips report displays in promoting the importance of non-expert knowledge causes the report to make a further mistake. Lord Phillips assumes that a greater understanding of government- sponsored scientific research will lead to the greater public acceptance of the uncertain processes through which risk is derived and managed. The potential for fostering trust through the promotion of public participation in the regulatory processes and the democratization of expertise more generally is largely ignored. If, as a society, we wish to be better able to contend with uncertainty and the potential for risk, then we need to begin to foster public inclusiveness, both in how we define and how we structure relations of knowledge and expertise (Irwin 1995; Wynne 1996). Colin Tudge cogently states the problem that Phillips has chosen to ignore:

> Lord Phillips might also have asked why we still tie ourselves so complacently to the establishment. The committee that met under Sir Richard Southwood in 1988 to assess the unfolding of BSE was learned, eminent and well-intentioned. Yet it left stones unturned. Thousands of ordinary, intelligent people, who were neither learned nor eminent, would have done a much better job. . . . They would have asked awkward questions such as 'how do you know?' and "why not?". (Tudge 2001: 27)

Conclusion

The Phillips report makes an excellent case for the need to incorporate sociological notions of risk into our understanding of the hazards and complexities involved in the BSE story. Clearly, governments need to move beyond expectations of scientific certainty in governing risks, and the practices of secrecy that characterized the BSE story. The uptake of conceptions of risk and uncertainty have certainly aided Phillips greatly in making this case. However, by enclosing notions of risk and uncertainty within a framework that takes modernity for granted and emphasizes notions of the inevitability of risk, the Phillips report is unable to produce a more radical social commentary on the BSE story. The report finds solutions to the BSE story in better science, practice and government. It ignores the issues of consent and the normative politics surrounding the use of intensive farming practices that not only provided the context in which the disease developed, but that also are at the heart of public reactions to BSE. In doing so, Phillips misses the opportunity to envision how the democratic inclusion of alternative forms of knowledge can respond to risk.

In conclusion, BSE provides an example of a situation in which sociological theories of risk are not only relevant, but essential, to the governance of modern agricultural practices. However, the sociology of risk must give greater consideration to how conceptions of risk are being taken up in government contexts. Theories of risk and uncertainty offer many benefits to governments in how they handle the rapid transformations in knowledge and practice that characterize modern society. However sociologists of risk must also become more aware of the limitations exposed in their theories when ideas of risk and uncertainty are applied in practice.

Notes

List of Acronyms
BSE—bovine spongiform encephalopathy or 'mad-cow' disease
CJD—Creutzfeldt-Jacobs disease
DH—Department of Health
MAFF—Ministry of Agriculture, Fisheries and Foods
SBO—Specified Bovine Offal
TSE—transmissible spongiform encephalopathy
vCJD—new variant Creutzfeldt-Jacobs disease

1. Considerable public concern in the United Kingdom has been expressed over the perceived relationship between the administration of the vaccine and the development of autism.
2. The Phillips report (2000b: xviii) now speculates that BSE probably first surfaced in the 1970s. Thus, although the official history of the disease begins in 1985, its origins are somewhat older. This implication of which is that BSE infected material would have entered the nation's food supply around

this same time, although it is not possible to speculate to what extent this was the case.

3. The government's position on the human health risks of BSE come largely from the findings of the Southwood Working Party—an ad hoc expert panel created in 1988 to advise ministers on the potential ramifications of BSE. The subsequent report of the Working Party stated that it was "most unlikely that BSE would have any implications for human health" (cited in Phillips et al. 2000b: xx).

4. The fact that Phillips adopts many of the tenets of the sociology of risk is not particularly surprising. The relationship between New Labour's presentation of Third Way politics and Giddens' proclamations about the nature of 'late modern' British society are clearly interwoven. Whether taken as the serious attempt by a government to better understand the social and economic relations that are shaping modern Britain, or as simply a tool with which to theoretically justify government decisions (Francis Wheen "Please, enough of the guff," *The Guardian*, 24 November 1999), risk theory has become diffused throughout the current political culture.

5. Despite making this acknowledgement of uncertainty, the report would appear to argue that the limitations of science are only temporary. In particular, the report cites recent developments in molecular biology as holding the key to providing the sense of scientific certainty that was so conspicuously absent during the BSE story. Termed the "rogue prion" hypothesis (Ridley and Baker, 1998, see also Prusiner 1999), it is argued that TSEs result from the presence of prions (transmitted or inherited) that force benign molecules to change their shape, thus converting normal proteins into dangerous ones. This thesis challenges many of the taken-for-granted assumptions about molecular biology, including the primacy of genetic material, such as DNA and RNA, in the transmission of diseases (Prusiner 1995: 531). Irrespective of this controversy, the report continues to look to the promise of factuality and the progress of science as the primary tools in contending with BSE and uncertainties of knowledge in the future (Phillips et al. 2000e).

6. This conflict of interest between protecting consumer and industry interests was one that was structurally imbedded MAFF, the lead agency in charge of the handling of BSE, was simultaneously the primary food safety regulator in the United Kingdom and tasked with the development of the agriculture and food industries. For a good discussion of this conflict mandate please refer to Little (2001).

7. Miller (1999: 1242–3) demonstrates that such failures are contextual. In particular, he reminds us that the BSE story took place during a time in which consecutive Conservative governments held power—governments that were characterized by deregulation, privatization, and a general "concerted tilt to the market in government policy." The emphasis of the government was not on safety, but on allowing industry to best determine how to increase productivity and profits. The use of ruminant feeds developed out of this context.

8. For a further discussion of BSE and the language of hysteria that characterized the government's relationship with the public please refer to the work of Bob Hodge and Robert Woog (1999).

9. One article in *The Guardian* titled "Fallibility in a White coat" (Hywel Williams, 23 January 2001) makes the provocative statement that "government grants and private sector research alike tie in the scientist to specific aims. Scientists are less free spirits than intellectual castrati singing for their *table d'hote* supper. Others reported on suggestions made in the Phillips report that scientific voices had allowed themselves to be muted by government and industry interests (John Crase "Silent Witness" 31 October 2000). In Britain scientific authority is now the subject of routine media and public scrutiny (Cristina Odone, "Science in the dock: The white coats are looking grubbier after one too many scandals," *The Observer*, 17 February 2002).

References

Beck, Ulrich (1992) *Risk Society: Towards a new modernity.* London: Sage Publications.

—— (1996) "Risk Society and the Provident State," in *Risk, Environment and Modernity: Towards a new ecology,* Scott Lash, Bronislaw Szerszynski, and Brian Wynne (eds.). London: Sage Publications. 27–43.

—— (1998) "Politics of Risk Society," in *The Politics of Risk Society,* Jane Franklin (ed.). Cambridge, U.K.: Polity Press. 9–22.

The BSE Inquiry: Statement No. 311 (1998) *Statement of John Selwyn Gummer MP. House of Commons, London SW1A 0AA,* December.

Douglas, Mary, and Aaron Wildavsky (1982) *Risk and Culture: An essay on the selection of technical and environmental dangers.* Berkely: University of California Press.

European Commission (1996) "Commission Decision 96/239/EC of 27 March 1996 on emergency methods to protect against bovine spongiform encephalopathy." *Official Journal of the European Communities* L78(28 March): 47.

Ford, Brian J. (1996) *BSE: The Facts—Mad cow disease and the risk to mankind.* London: Corgi Books.

Guldens, Anthony (1990) *The Consequences of Modernity.* Cambridge, U.K.: Polity Press.

—— (1994) "Living in a Post-Traditional Society," in *Reflexive Modernization: Politics, tradition and aesthetics in the modern social order,* Ulrich Beck, Anthony Giddens, and Scott Lash (eds.). Cambridge, U.K.: Polity Press 56–109.

—— (1998) "Risk Society: The Context of British Politics," in *The Politics of Risk Society,* Jane Franklin (ed.). Cambridge, U.K.: Polity Press. 23–34.

Grove-White, Robin (2001) "New wine, old bottles? Personal reflections on the new biotechnology commissions," *The Political Quarterly* 72(4) 466–72.

Hodge, Bob and Robert Woog (1999) "Beyond reason in hysteria: Toward a postmodern model of communication and control in science." *Social Semiotics* 9(3): 375–92.

Irwin, Alan (1995) *Citizen Science: A study of people, expertise and sustainable development.* London: Routledge.

Irwin, Alan, Alison Dale, and Denis Smith (1996) "Science and Hell's Kitchen: The local understanding of hazard issues," in *Misunderstanding Science? The public reconstruction of science and technology,* Alan Irwin and Brian Wynne (eds.). Cambridge: Cambridge University Press. 47–64.

Jones, Kevin E. (2001) "BSE, risk and the communication of uncertainty. A review of Lord Phillips' report from the BSE Inquiry," *Canadian Journal of Sociology* 26(4) 655–67.

Kitzinger, Jenny, and Jackie Reilly (1997) "The rise and fall of risk reporting Media coverage of human genetics research, 'false memory syndrome,' and 'mad cow disease,'" *European Journal of Communication* 12(3): 319–50.

Latour, Bruno (1987) *Science in Action: How to follow scientists and engineers through society*. Cambridge: Harvard University Press.

Leach, Joan (1998) "Madness, Metaphors and Miscommunication: The Rhetorical Life of Mad Cow Disease," in *The Mad Cow Crisis: Health and the Public Good*. Scott C. Ratzan (ed.). London: UCL Press. 119–30.

Little, Gavin (2001) "BSE and the regulation of risk." *The Modern Law Review* 64(5): 730–56.

Michael, Mike (1996) "Ignoring Science: Discourses of Ignorance in the Public Understanding of Science," in *Misunderstanding Science? The public reconstruction of science and technology*, Alan Irwin and Brian Wynne (eds.). Cambridge: Cambridge University Press. 107–25.

Miller, David (1999) "Risk, science, and policy: Definitional struggles, information management, the media, and BSE," *Social Science and Medicine* 49: 1239–55.

Millstone, Erik, and Patrick van Zwanenberg (2001) "Politics of expert advice: Lessons from the early history of the BSE saga." *Science and Public Policy* 28(2): 99–112.

Murphy, Raymond (2001) "Nature's temporalities and the manufacture of vulnerability: A study of a sudden disaster with implications for creeping ones," *Time and Society* 10(2/3): 329–48.

Narang, Harash (1997) *The Link: From sheep to cow to man. Creutzfeldt-Jakob disease, bovine spongiform encephalopathy, scrapie*. Newcastle-upon-Tyne: H.H Publisher.

Perrow, Charles (1984) *Normal Accidents: Living with high-risk technologies*. New York: Basic Books.

Phillips, Lord of Worth Matravers, June Bridgeman, and Malcolm Ferguson-Smith (2000a) *The BSE Inquiry Report: evidence and supporting papers of the inquiry into the emergence and identification of bovine spongiform encephalopathy (BSE) and variant Creutzfeldt-Jakob Disease (vCJD) and the action taken in response to it up to 20 March 1996*. London: The Stationary Office.

—— (2000b) *The BSE Inquiry—Volume 1, Findings and Conclusions*. London. The Stationary Office.

—— (2000c) *The BSE Inquiry—Volume 4, The Southwood Working Party 1998–1999*. London: The Stationary Office.

—— (2000d) *The BSE Inquiry—Volume 8, Variant CJD*. London: The Stationary Office.

—— (2000e) *The BSE Inquiry—Volume 2, Science*. London: The Stationary Office.

—— (2000f) *The BSE Inquiry—Volume 5, Animal Health 1989–1996*. London: The Stationary Office.

———— (2000g) *The BSE Inquiry—Volume 11, Scientists after Southwood.* London: The Stationary Office.

———— (2000h) *The BSE Inquiry—Volume 13, Industry Processes and Controls.* London: The Stationary Office.

Prusiner, Stanley B. (1999) *Prion Biology and Diseases.* Cold Spring Harbor. N.Y.: CSHL Press.

———— (1995) "The prion diseases," *Scientific American* 272(1): 530–31.

Ridley, Rosalind M., and Harry F. Baker (1998) *Fatal Protein: the story of CJD, BSE and other prion diseases.* Oxford: Oxford University Press.

Southwood, Sir Richard, M.A. Epstein, W.B. Martin, and J. Walton (1989) *Report of the working party on Bovine Spongiform Encephalopathy.* Department of Health, Ministry of Agriculture, Fisheries and Food, United Kingdom London: The Stationary Office.

Tudge, Colin (2001) "Bring back common sense," *New Statesman* 14(641, January 29): 25–7.

Woods, M. (1998) "Mad cows and hounded deer: Political representations of animals in the British countryside," *Environment and Planning A* 30: 1219–34.

Wynne, Brian (1996) "May the Sheep Safely Graze? A reflexive view of the expert-lay knowledge divide," in *Risk, Environment & Modernity: Towards a new ecology,* Scott Lash, Bronislaw Szerszynski, and Brian Wynne (eds.). London: Sage Publications. 44–83.

9

Subjects of Knowledge: Epistemologies of the Consumer in the GM Food Debate

Javier Lezaun

After announcing their visit two days in advance, on 27 March 1999 four members of the "genetiX snowball" campaign walked into a Tesco supermarket in London and "confiscated" several baskets full of products allegedly containing genetically modified ingredients. The four visitors refused to pay for them, but offered to exchange them for organic products. One of the activists was consequently arrested for theft. As she was led away by the police, she said, "Tescos are breaking the law by selling food which is not proven to be safe and which is endangering other farmers' crops in the production process through genetic pollution. I intend to carry on decontaminating supermarkets and I hope others will join in."[1] Two days later, the London "snowballers" carried the confiscated products in a biohazard dump to the Ministry of Agriculture, Fisheries and Food, and asked the Ministry officials to take responsibility for the safe disposal of the foods. In written statements the activists described the "decontamination" as an "act of nonviolent civil responsibility," and pledged to carry out similar actions in the near future.

GenetiX snowball initiated its campaign against genetically modified organisms on 4 July 1998, with the pulling up of several dozen plants at the Model Farm of the U.S. biotechnology company Monsanto, in Watlington, Oxfordshire. As in the Tesco action, the activists openly declared their intentions in advance in letters to Monsanto, the police, and the farmers hosting test field sites for genetically modified crops in the area. Five women were arrested and later released that day, and many other people were to be arrested in subsequent actions throughout England. Borrowing tactics from the peace movement of the 1980s, in particular the ploughshares actions of the previous decade, the snowballers targeted demonstration fields and test sites where genetically modified crops were being grown and pulled up a small number of plants. GenetiX snowball was one of many groups involved in the sabotage of genetically modified crops, which in 1998 became a highly visible form of protest against genetically modified organisms in the United Kingdom. What distinguished their actions from those of other groups was a special commitment to openness and the personal acceptance of legal consequences. The "ground rules" of the campaign were laid out in a *Handbook for Action* (1998), but the particular tactics and their effectiveness were often an object of discussion and disagreement among activists, in the face of severe legal challenges from the biotechnology companies whose test fields had been affected (genetiX snowball 1998).

The London action, however, marked a shift in the tactics of the snowballers. For the first time, a supermarket, not plants in a test plot, was being targeted. The campaign moved thus from the country to the city, from the decontamination of fields to the confiscation and disposal of hazardous consumer products, assuming the principle that, if the release of genetically modified organisms had to be prevented, their market release had to be stopped as well. Thus, after several months of resistance against genetically modified plants, the campaigners decided to extend the reach of their actions. "So far, genetiX snowball has focused on the production end of GM food—the GM crops in our fields," an article on the campaign's website pointed out "But is it really enough to keep the genetic peril from our own field whilst it is being imported from fields in other countries? We also need to pay attention to the consumer end of things—the GM products in our supermarkets."[2]

The case of genetiX snowball is only one example among many others in the shift in the tactics of the resistance to genetically modified organisms in the United Kingdom and, more generally, in Europe. Large environmental campaign organizations like Friends of the Earth and Greenpeace, had come to believe that the battle for the defense of the environment had to be fought also, if not primarily, in the supermarket and the grocery store. The *locus* of resistance therefore shifted to the practices of everyday consumption and the defense of "consumer rights." In an often-heard expression. Friends of the Earth described the introduction of genetically modified foods in the United Kingdom as "a crime against consumer choice." Friends of the Earth initiated in 1998 their 'Supermarket Challenge' campaign, proclaiming that that "as a consumer, you have impressive power!" Greenpeace published a popular "supermarket shoppers' guide," to help consumers avoid products likely containing GM ingredients. At the same time, several British newspapers carried regular sections advising readers on how to avoid consuming GM foods.[3] The result is well-known: by the spring of 1999, all major supermarket chains and food manufacturers had vowed to eliminate genetically modified ingredients from their products.

The example of genetiX snowball is significant precisely because their previous direct actions were deeply rooted in the traditions and tactics of the ecological and peace movements, and in that sense represented an extreme in the spectrum of groups and organizations opposed to the release of genetically modified organisms. But now even they were adopting the tactics of the more mainstream organizations. From an initial emphasis on the protection of the environment, the center of gravity of the campaign had moved to consumers and forms of protest that targeted primarily food retailers and producers. In fact, the campaign against supermarkets was so successful that, by the time the snowballers staged their London action, only two supermarket chains—Tesco and Safeway—were still refusing to remove genetically modified ingredients from their own brand products. And a few weeks after the snowball action. Tesco, the UK's biggest supermarket chain, decided to eliminate these ingredients altogether, and announced that it was collaborating with Greenpeace to find non-GMO crops around the world. It was clear that it was in the marketplace, and in the name of the consumer, where the battle for the success or failure of genetically modified foods was to be fought out.

Consuming Genetically Modified
Foods—From the Industry's Point of View

This essay addresses some of the processes through which the consumer of the GM food debate was constituted. In a way it explores an overlooked prehistory of these events described in the introduction. Prehistory in the sense that it explores events and positions that took place or were decisively shaped, for the most part, *before* the intense public debate and widespread mobilization against genetically modified foods, and in the absence of broad and articulated consumer resistance. The essay, moreover, does not analyze opposition to genetically modified foods, but the other side of the coin: the strategies developed by the British food industry since the early 1990s to ensure a successful marketing of genetically modified foods and to guarantee consumer acceptance of the new technology. More specifically, it will describe a particular understanding of the consumer produced by the British food industry through a variety of consumer research tools, an understanding that was central to the efforts of the industry to regulate itself. The image of the consumer that the industry developed was strikingly similar to the subject of opposition to genetically modified foods that environmental groups were to mobilize in the late 1990s. A very specific discourse about how to understand the consumer, the role of choice in the governance of biotechnology, and the centrality of information was common to both opponents and proponents of bioengineered foods. This essay describes how the industry reached this understanding as part of its own commercialization strategy, and how the strategy then unraveled in the face of actual consumer pressure. I will then draw some conclusions as to the governance of markets and technologies through the mobilization of consumers.

The United Kingdom presents a number of interesting variables to any study of consumer activism and public understanding of science in the 1990s. The country at the time was shattered by a number of scares and regulatory crises in the area of food safely and biotechnology. BSE (or "mad cow disease") was of course the main episode in this chronology, and Kevin Jones analyzes the institutional response to the crisis in Chapter 8 of this volume. A deficit of institutional legitimacy affecting most regulatory institutions lies at the heart of the story told below.

Regarding bioengineered foods and genetically modified organisms, the British food industry provides us with a useful case study. It can be argued that the United Kingdom was the first market where genetically modified foods failed, as consumer products, on a massive scale. Even though opposition to genetically modified organisms was already strong in other European countries, the explosion of public debate and consumer resistance took place in the United Kingdom at a critical moment in the commercialization cycle of the new products: from late 1997 to mid 1999, after the first shipments of genetically modified crops had arrived in Europe, and when food manufacturers and retailers were confronted, for the first time, with unmanageable amounts of genetically modified ingredients in the food supply. The likely presence of GMOs in food products and grocery stores provided the different mobilization strategies with a new, practical focus, and opened the possibility of new forms of action—boycotts, campaigns targeting food retailers, "supermarket decontamination," etc.

The British food industry was, however, not unprepared for the arrival of genetically modified foods. In fact, it had long tried to develop a common, "food chain" approach to the technology, a common plan for the commercialization of the yet-to-arrive food products, and a pre-emptive communication strategy to allay public fears. The initial success of the effort is often forgotten, in light of the ultimate failure of its strategy in 1998–9. A clearly labeled genetically modified food was successfully commercialized in the United Kingdom in 1996, barely two years before the onslaught of public opposition. The product was a tomato purée produced from genetically modified tomatoes developed by the British company Zeneca. The tomato was commercialized under their own brands by two large British supermarket chains, Sainsbury's and Safeway, in February of 1996. The cans of tomato purée were clearly labeled as "produced from genetically modified tomato," and were sold at a cheaper price than conventional varieties (Harvey 1999).[4] The launch of the product was extensively publicized: according to the industry's own data, the televised launch of this biotechnology product reached approximately 22 million viewers in the United Kingdom. Sainsbury's and Safeway sold 1.6 million cans.

Most of the activities described in the following sections can be described as "consumer research," a type of knowledge production that has received scant attention from sociologists and historians of knowledge. By 'consumer research' we often refer to a variety of techniques of investigation and representation—public polls, focus groups, media analysis—and a variety of experts and expertise—sociological, psychological, statistical, managerial—aimed at producing a true and operational depiction of the consumer of a particular product, the 'subject of consumption.'[5] This knowledge is produced by or on behalf of corporations, and is intended to serve their own interests. It is a private knowledge of the public that, insofar as it is produced and acted upon by economic actors, remains internal to the constitution of the marketplace. Indeed, consumer research techniques are tools of knowledge production that "perform the economy."[6] In his analysis of "the laws of the market," Michel Callon (1998) argues that marketing and accounting tools constitute, and continually disrupt, the balances of the economy.

Similarly, in our case, a particular image of the subject of consumption, produced through consumer research, is *both* a way of managing the market *and* a potentially disruptive element in the calculations and actions of economic actors.

All the actors and corporations involved in the making and marketing of bioengineered foods conducted their own consumer research in an effort to map out the expectations, needs, attitudes, and future reactions of consumers. Large food companies and biotechnology firms earned out their own studies from very early on and tried to assess future market trends and patterns of consumer behavior.[7] Much of this information is proprietary and often inaccessible to outside researchers—an indication of the value companies attribute to it Glimpses of this vast body of knowledge about subjects and practices of consumption appear in public speeches by company executives, or can be deduced from marketing and communication strategies. But the specific processes of production, interpretation and deployment of such knowledge are frequently opaque to the public, and the outside researcher in particular. This essay is thus limited to the study of the body of knowledge and policy produced by an industry actor with an unusually high public profile: the Institute of Grocery Distribution (IGD). The IGD is, nevertheless, an important and representative agent: it is the research arm of the British Food and Drink Federation and the British Retail Consortium, which represent, respectively, food manufacturers and supermarket chains operating in the United Kingdom. The IGD produces research for the British food suppliers and is central to the ongoing effort to unify the actions, strategies and interests of the different actors of the food chain, from farmers to retailers.

Though the IGD, the industry produced a common understanding of the consumer of biotechnology foods—of consumers' fears and interests, abilities and rights to know, and potential for resistance. This understanding was based on the notion of informed consumer choice and respect for consumer rights as the key to gaining public acceptance for the new technology. If consumers are given a "free and informed" choice in the marketplace, the argument goes, the technology will be generally accepted, and public concerns will be mitigated. This understanding helped produce a commercial strategy to find "the path of least resistance" to the British market, a strategy that was exemplified in the communication and labeling guidelines that the IGD issued in March 1997.

Consumer Research and Self-Regulation

The British food industry claimed to have learned the lessons of several crises in the late 1980s and early 1990s, particularly the opposition to food irradiation. This time it decided to act pre-emptively on a technology, namely biotechnology, whose future impact was clear. In 1994, the Policy Issues Council of the food industry, formed in 1991 as part of the IGD, established a so-called Biotechnology Initiative. The goal was to understand "consumer attitudes and consumer requirements," and to identify both a "strategy for the introduction of products of biotechnology in order of consumer acceptance," and the "retailers and manufacturers with the customer profile most likely to accept the new technology." As Sir Alistair Grant, then-president of the IGD, pointed out, "the success of biotechnology in food provision is clearly dependent on consumer acceptance of the products. This will depend on the use of appropriate information and careful introduction of products."

The IGD established a Biotechnology Advisory Working Group in 1994, composed of representatives from the largest supermarket chains, from large food manufacturers like Unilever and Nestlé, a biotechnology company (Zeneca), the National Farmers Union, and the Consumers' Union. After several months of discussion among these stakeholders, the Working Group issued the first statement on the commercialization of food biotechnology in October of 1995, which was then endorsed by the Policy Issues Council. At that point, no foods containing genetically modified ingredients had been commercialized in the United Kingdom. In the statement the Working Group claimed that:

> As an industry we are committed to a policy of openness and facilitating understanding as a means of addressing any concerns about the new technology. We believe that the provision of information is essential to enable customers to make an informed choice about food products. The industry will endeavour to make information available in the most effective manner to give an objective and balanced view of genetic modification.[8]

The central concepts that would inform the future public initiatives and communication strategics of the industry are already present in this statement: "openness" (or later, "transparency") and, more crucially, "informed choice." No explicit reference to labeling was included in the position, but by the end of 1995 this was the central issue being discussed by the industry representatives. Supermarket chains were particularly keen on labeling, and saw a "practical and meaningful" identification of the products as the key to gaining public acceptance of the technology. Large food manufacturers like Unilever and Nestlé were moderately opposed to strict labeling requirements, while biotechnology companies and especially providers of raw materials (like Cargill) were radically opposed to the idea.

It was in the context of these disagreements that the IGD initiated its consumer research activities. They organized a number of focus groups and surveys, and began to develop a parallel communication strategy, based on consultation with consumer groups and activists (Sadler 2000). The image of the consumer that, according to IGD researchers, consistently emerged from this research was one of individuals with low level of awareness and understanding of the issue, but with a very strong demand for "information". It was assumed that the path to consumer acceptance passed through a labeling policy that recognized desire for 'choice.'

But how much did the biotechnology consumer want to know? If the label on the food product materialized the right to know and the need for informed choice, it was still necessary to determine which products—of the many containing or produced from genetically modified organisms—were to be labeled and how. Again, the industry undertook a program of qualitative research to explore consumers' reactions to GMO labeling. The question the industry wanted to answer was whether following the existing labeling regulations would be enough to defuse potential consumer concerns over biotechnology. According to these regulations, still in the making at that time, food products were to be labeled only if they contained intact modified DNA. But, based on the results of several focus groups, the industry concluded that this would not be enough to allay future public concerns. As an IGD report argued,

> The important issue for consumers would be the presence of *modified genetic material*, a novel entity that would be perceived to present a potential risk. . . . Consumers would be unable to differentiate between genetic material that is viable, (intact; active) and non-viable (degraded through processing: inactive). (Sadler 2000: 45.)

The success of Zeneca's tomato purée in 1996 helped consolidate a theory of consumer behavior that linked acceptance of the new technology to the provision of clear labels on the products. The success of the tomato purée was a result, the industry interpreted, of the intense communication strategy (the more publicity, the better), the protection of consumer choice through labeling, and the provision of a direct consumer benefit (a competitive price). It corresponded perfectly to the strategy devised by the industry. As one IGD official put it, "in effect, the group had adopted a strategy of step-by-step introduction of GM foods, isolated, one at a time; with, as far as possible, some definable consumer benefits; with labeling and information. This [the Zeneca experience] was doing it by the book."[9] The industry "recognised the consumers' desire for choice and for labeling," and the IGD drew the conclusion that "public confidence would be gained by a clear industry policy and appropriate labelling" (Sadler 2000: 19). The strong link between labeling and consumer acceptance of the technology had become evident to the eyes of industry researchers.

This is not to mean that the results of the IGD's consumer research and the lessons of Zeneca's tomato puree were interpreted in the same way by all the industry actors. It is true, however, that for the researchers at the IGD, the results of the focus groups and of qualitative surveys were essential in bringing the food industry itself to understand the importance of labeling and consumer choice; they were an instrument for changing perceptions within the industry. As one of the IGD members involved in the "consumer priorities program" points out,

> Whenever you have a group of stakeholders around, from the food chain, they would come with different perceptions, different positions, and mindsets. But they all understand the commercial benefits of pleasing consumers, and the commercial risks of not doing so. And, therefore [consumer research] is like a trump card, as a consensus builder. It is the trump card that IGD can always play.[10]

But, and as one would expect, the specific results of the research were often controversial among the members of the Biotechnology Advisory Working Group. In particular the possibility of drawing general conclusions about 'the consumer' from a limited population of focus group participants was always a potential target of criticism. The methodology was questioned even further when it became known that some of the participants in the focus groups were members of consumer organizations—and very vocal members at that. It seems that, at critical moments, the abstract and generalized consumer of the focus groups became increasingly difficult to sustain, increasingly specific, and therefore unable to represent or stand for the consumer public as a whole.

Most of the conclusions that IGD researchers drew from the focus groups resonated with the corporate philosophy of several major players in the UK food manufacturing and retail industries. Many retailers, in particular, had built their corporate identity on a managerial practice based on "listening to the consumer," or had created "consumer insight" departments, etc., and the conclusions of the IGD consumer research programs echoed many of these deeply ingrained views of the marketplace. Besides, the idea of subordinating the commercial strategy to the principle of consumer choice was appealing to economic actors who had, at this stage, no direct investment or interest in the success of specific bioengineered products.

Finally, after more than a year of internal deliberations, in March 1997 the IGD Policy Issues Council issued a series of specific recommendations concerning the labeling of products containing genetically modified material and the communication strategy that the food industry should adopt vis-à-vis the consumer. This was the outcome of the industry's effort to come up with its own regulations, and to do so in a concerted and coherent manner, while avoiding divisive competitive strategies. Ross Buckland, chief executive of Unigate and IGD president, urged the food industry to adopt the regulations in order to "demonstrate a positive commitment to consumer understanding and choice."[11] The guidelines were the result of multiple meetings, disagreements and conflicts of interest among industry sectors and individual companies, but these appear nowhere in the final document. A certain shared understanding of the situation facing the industry had been produced, and the lessons of the past had been interpreted in a more or less homogeneous way. "As we have seen with food irradiation," the guidelines document reminded its readers, "new technologies are not always readily accepted by consumer. The provision of freely available, objective information and, where practicable, informed choice are key to the successes of these products" (Institute of Grocery Distribution 1997: 18).

The guidelines were indeed a remarkable exercise in self-regulation. They were issued at a time when, according to most polls, the public was still showing low awareness of the issue of biotechnology foods. The labeling guidelines, in particular, were stricter than either the British regulations then in place, or the legislation under preparation by the European Commission— what later became the European Union Novel Foods Regulation. To meet what the industry perceived as a strong demand for informed consumer choice, the recommendations called for the labeling of *any* foods *known* to contain modified DNA or protein. Regardless of whether they were substantially equivalent to conventional foods or not, and whether the modified DNA or protein was intact or not, a food "should be labelled if it is known to contain modified genetic material, whether active or not" (Institute of Grocery Distribution 1997: 7).[12] This all or nothing approach conformed to the mainstream consumer's alleged inability to distinguish between different types or degrees of modification.

In its proposed regulation, the IGD in fact anticipated many of the changes in labeling provisions that were to materialize in future European regulations by abandoning the (by now forgotten) principle of "substantial equivalence" (that is, labeling only those products that are substantially different from their conventional counterparts), and the principle detectability (labeling only those products that contain detectable DNA or protein that can be detected with the existing testing methods.) They instead adopted a methodology of labeling based on the ability to know whether a certain product contains DNA or protein (any product "known to contain modified genetic material"), in whatever form or at whatever level—a notion that prefigures current provisions based on 'traceability.'

Market Failure, Culprits, and Assumptions

How the story developed from this point on is well known. What had been a concerted strategy by different sectors in the food chain turned, under increasing consumer pressure, into a series of individual, uncoordinated actions by the different sectors and companies within sectors as they tried to limit the impact of the public controversy on their brands. The IGD had developed its consumer and labeling guidelines in a context lacking direct public participation m the debate. But a few months before the IGD released the labeling guidelines, the first shipments of genetically modified soybeans had arrived in European ports, and an intense campaign by environmental and consumer groups ensued. The themes of the campaign resonated profoundly with the notions and principles that underpinned the industry's discourse: the consumer's right to know, the right to an informed choice, the importance of labeling, etc. Soon, environmental groups would be drawing on these notions, trying to shape "the consumer end of things." They were mobilizing the same consumer that the industry had tried to understand and represent, *against* the commercialization of genetically modified foods. Much of the British media joined in on the mobilization of consumer resistance. An editorial in *The Express* argued that "with the Government determined to enable the continuing development of products nobody seems to want, it is up to consumers to press for a real choice—and to pressurise all food retailers to provide it."[13]

By the spring of 1999 the GMO issue had become, according to the industry's own polls, the main "food safety concern" of British consumers, surpassing pesticides, food poisoning and even BSE. Beginning in 1998, some supermarket chains had begun to offer "GMO-free" products, or promised to phase out genetically modified ingredients from their own-brand products.[14] Iceland, a food retail chain that was not part of the IGD Biotechnology Advisory Working Group and commanded seven percent of the market in the UK, was the first to publicize its products as "containing no GMOs." Very soon, all the major supermarket chains, and the largest food manufacturers had announced that they would try to avoid using genetically modified ingredients in their products. After Tesco and Unilever announced the phasing out of genetically modified ingredients from their products in April 1999, *The Independent* published, under the title "Consumer Victory," a triumphant tribute to consumer power:

> What a good week this has been for those who believe in the power of the consumer. Nothing, we had been told, was to stand m the way of the progress that was genetically modified food; only Luddites and hysterics, we were led to understand, had doubts about health implications; why wait for further testing, said those who know better, when the technology was available now? The consumers didn't accept any of this, and made it clear that they wanted more information before buying new foods. One by one the supermarkets, which had started selling genetically modified products without so much as a blush, began to change their tune.[15]

Environmental campaign organizations and the media were operating in a vacuum of governmental regulation, in the absence of a forceful and coherent governmental response. And in a post-BSE environment, no food company could afford losing the confidence of its customers. One after another, supermarket chains abandoned the common IGD position, and adopted "no GMO" policies of their own. As the trade journal *The Grocer* pointed out, "consumers are now looking to the supermarkets instead of the government to take the moral and defensive stance on their behalf. And from a commercial point of view, the issue about GM products is also about managing fear successfully and consequently maintaining or, in the case of Iceland, even increasing market share."

The public reaction took the IGD by surprise. Their studies consistently showed a "low level of awareness" among consumers, and the "successful" commercialization of the Zeneca tomato purée still dominated the industry's understanding of the British consumer. To make sense of this new reality, the IGD commissioned a national survey in December of 1998, and a new round of qualitative consumer research in March of 1999. The researchers concluded that there had been no significant change in the makeup of the British consumer: "the main effect of the storm—the report concluded—was to raise awareness of GM foods, rather than to switch consumers from the unconcerned to the concerned category."

The IGD also tried to find culprits for the failure of its efforts to manage consumer concerns. The industry blamed the media for its coverage of the GMO issue. Irrational fear fueled by irresponsible media coverage had replaced the consumer's legitimate and measured demand for "meaningful" information, the IGD argued. In this new environment, the industry's communication and labeling guidelines made little sense:

> The original industry strategy of choice, labelling and information had become unsustainable in the wake of the exceptional public reaction. No business could run the risk of maintaining GM ingredients in their products (Sadler 2000. 62).

The introduction of non-segregated, unlabeled modified soybeans from North America made these IGD labeling recommendations technically impractical, companies had no way of knowing whether the ingredients of their products were conventional or modified. And different soy derivatives and ingredients are present in a multitude of processed foods. The culprit in this case was easy to identify: the American company Monsanto. The retail sector had tried to convince Monsanto, not a participant in the IGD "food chain" efforts, to segregate production of its herbicide tolerant soybeans in the United States, but to no avail. When the public's opposition to genetically modified products became clear, the food industry was eager to blame the failure of its effort on Monsanto's uncooperative stance. The introduction of Monsanto's soybeans seemed to violate all the truisms of the IGD's understanding of the consumer of bioengineered foods.

> The arrival of Round-up Ready [genetically modified, herbicide tolerant] soya in the UK was a crucial turning point for consumer

acceptance of GM foods. The inclusion of a GM soya variety in the commodity stream was in contrast to the UK industry's desired approach to introducing GM products. GM soya had no direct consumer benefit, and without segregation, consumers would not easily be able to exercise choice. With soya ingredients used in an estimated 60% of processed foods, the possible presence of GM soya ingredients in a wide range of products would give consumers the impression of a very fast introduction of the technology. (Sadler 2000: 27.)

The industry had defined though the IGD Biotechnology Advisory Working Group a scenario in which the consumer would be able to choose between clearly labeled genetically modified products and conventional ones. In this scenario, the industry had concluded, consumers would become increasingly accustomed to the presence of bioengineered foods, and would eventually discover, in a next generation of products, the benefits of the technology. Labels were the way of providing choice and information. This scenario has not materialized. To this day, no labeled products can be found in British, or for that matter, European supermarkets; meanwhile, manufacturers and retailers try to source raw materials free of genetic modification.

Since the IGD developed its particular understanding of the consumer of biotechnology, some in the industry have moved towards a new notion of the consumer, and of what it takes to commercialize a controversial technology. The model of the "epistemic" consumer, an actor whose competencies and behavior are defined in terms of "understanding" the issue (or, more frequently, *not* understanding it), and whose "right to know" is satisfied through informed choice in the marketplace, might be giving way to a new image of the subject of consumption. Perhaps the approach should be different, a recent IGD report argues, if the key issue for consumers would be "a lack of control, rather than low understanding as it is often assumed" (Sadler 2000: 81).

Moving away from an informational notion of choice to an emphasis on control opens up intriguing possibilities. In his analysis of cultural and sociological terms, Raymond Williams found notion of the "consumer choice" paradoxical (Williams 1983 [1976]: 78–9). It was a "curious phrase," he argued, because, historically, the term 'consumer' was a product of the age of mass production and increased corporate control over the market. The notion of 'the consumer' implied an abstract actor, operating in an abstract market, and defined solely by the act of consumption, while 'customer' used to denote a personalized and regular relationship between the buyer and the seller. Consumer choice is a somewhat paradoxical concept because the emergence of the notion of 'the consumer' went hand in hand with the individual's loss of actual control over market forces and exchanges.

Conclusion: Mobilizing the Consumer

How should we interpret the IGD's consumer research and policy recommendations in the context of the GM food debate? How should we analyze consumer research as a knowledge-producing enterprise? To some observers, consumer research is partly an instrument to produce knowledge about the public, and partly a public relations strategy. Historian Roland Marchand has pointed out that the development of consumer research by companies like General Motors in the 1930s was designed primarily as an advertising tool, under the excuse of requesting the consumer's opinion on different products and brands (Marchand 1998). Much of this is true for the case of the IGD. The communication and labeling guidelines are, in and of themselves, part of the communication strategy, a marketing tool of the industry as a whole. Hence the high public profile of the IGD's projects and policy proposals.

In the case of the IGD, there is an additional consideration in interpreting the focus groups and other market research tools that were used to "listen to the consumer." In a context of uncertainty as to the future public's reaction to biotechnology in the United Kingdom, and industry's different view's on the matter, consumer research was intended as a consensus-building instrument among different industry groups. This was particularly important in the absence of precise governmental regulation on the labeling of bioengineered foods. Consumer research was to be the basis for the industry's self-regulatory strategy, a central component of the industry's attempt to understand the issues at hand, and govern itself. All the elements of the IGD's communication and labeling guidelines were justified on the basis of specific results of focus groups, surveys, and other market research instruments.

Moreover, far from being specific to the food industry, this particular understanding of the market structure and public attitudes towards biotechnology has since become central to the policy-making efforts of other European actors, to the extent of becoming a sort of common wisdom among regulators and politicians. A cursory analysis of recent speeches and position papers by European Commission officials and Directorates on biotechnology, for instance, demonstrates how widespread is the idea of free and informed consumer choice as the regulatory *panacea*. In presenting the new labeling and traceability

213

regulations for genetically modified organisms in July 2001. European Commissioner for Health and Consumer Protection David Byrne stated the goal and assumptions of this new regulatory approach: "The objective of the harmonised and comprehensive labelling requirements proposed," he argued, "is to respond to an overwhelming need to enable the consumer or users to make an individual choice and thereby to foster increased public confidence." The logic of this statement—informed choice in the marketplace leads to public confidence in a new technology, and in the institutions responsible for its regulation—could have been taken from any industry position paper.

If we understand consumer research as a way of producing knowledge about the public, knowledge that is then internalized and deployed by market actors, the actions of the IGD and of the British food industry can be described as an attempt to "mobilize the consumer"—in the sense in which Miller and Rose (1997) have used the term. As they argue—their study of advertisement in the 1950s, representing the consumer is a "complex technical process," "less a matter of dominating or manipulating consumers than of 'mobilizing' them by forming connections between human passions, hopes and anxieties, and very specific features of goods enmeshed in particular consumption practices" (Miller and Rose 1997).

Any mobilization implies the reduction and subordination of a complex situation to a particular strategy. The key question is not whether a true and objective account of what 'the consumer' is really like can be produced, for such an abstract consumer can only be the result of the techniques of investigation that actors like the IGD deploy. Mobilizing the consumer is not primarily a matter of getting it, the consumer, right (or, in this case in particular, wrong). Consumer research is an instrument for representing the world and intervening in it, a way of framing a complex reality to make it amenable to intervention. Let's not forget that the effort of the industry to understand the consumer had as its ultimate goal to find "the path of least resistance" to public acceptance of the new technology.

The interesting point is how a tool used to frame and manage the public, a discourse of consumer rights and informed choice that was intended to defuse irrational fears and limit the range of opposition, turned into a conduit for impossible demands (i.e., labeling in the absence of segregation), and a powerful resource in the hands of the opponents of biotechnology. The "consumer" can undoubtedly be an unruly subject. An analysis of market governance, or of consumer resistance to new technologies, should pay close attention to the concrete ways in which this abstract actor is mobilized, made compliant or unruly, and to the practices and techniques of knowledge production on which this mobilization is founded.

Notes

1. Genetix Snowball, Press Release, 27 March 1999.
2. GenetiX snowball. "The principles for supermarket decontamination." At http://www.gn.apc.org/pmhp/gs/shopping.html, retrieved 20 August 2001.
3. It is important to note here, even though the issue is beyond the scope of this essay, that the British environmental movement has a long and vexed relationship with the notion of "green consumerism," or the mobilization of citizens *qua* consumers as the main carrier of environmental demands. In the landmark book *Consumers' Guide to the Protection of the Environment*, first published in 1971 by Friends of the Earth, Jonathan Holliman argued that "the conversion to a life style more related to the ability of the Earth to supply our needs must start by the consumer regaining the political power of the individual to have real choice in the market place." Green consumer guides were extremely successful in the late 1980s. Some have argued that green consumerism was a "compromised response" to the philosophy of Thatcherism, a sort of hybrid between environmental activism and the free-market ideology that dominated British policy and politics in the 1980s and 1990s. The reliance on "consumer power" would also be a result of the difficulty of shaping the British policy process through direct, representative politics. The role of the state, as a target and potential ally of environmental activism would then become secondary, and a new emphasis is put on the ability of individual citizens to affect, in their capacity as consumers, changes in the economic system. See Robinson (1992).
4. As in the case of Calgene's commercialization of the FlavrSavr tomato in the United States, Zeneca and the British supermarkets thought at the time that publicity and open discussion of their respective products could only help their commercial prospects. See a first-hand account of the regulatory disputes involved in the FlavrSavr case in Martineau (2001).
5. This is similar to the survey efforts undertaken by governments or public agencies. For a study of the politics of (mostly) public polls and quantitative research on "the public acceptance of biotechnology," see Hornig Priest (2000).
6. Gallon (1998) argues, "The product is therefore a multidimensional reality, an entanglement of properties that the marketing mix disentangles," and the same could be said of 'the consumer.' He goes on: "The tool thus facilitates a more detailed analysis of buying decisions, as well as the preferences which they express or reveal. ... By enhancing the inventory of relations and events to be taken into account, marketing tools promote calculations which constantly involve more and more elements and relations. The formulation of these instruments which substantially increase the ability of producers and sellers to frame and internalize consumers and their preferences, helps to disrupt event trading practices."
7. A particularly good example is the study sponsored by Unilever and conducted by the Centre for the Study of Environmental Change, at Lancaster University. See Grove-White et al (1997).

8. IGD/PIC Biotechnology Advisory Working Group, October 1995.
9. Interview with IGD manager, January 2002.
10. Interview with IGD manager, January 2002.
11. Ross Buckland, Chief Executive of Unigate PLC, Chairman of the Policy Issues Council and IGD President, March 1997.
12. "Our discussions with consumers on this subject demonstrated that they were unable to distinguish between active and inactive (as a result of processing) genetic material" (ibid.: 5).
13. *The Express*, 28 April 1999.
14. Thereby violating the IGD guidelines, which urged that "under no circumstances should negative claims, such as 'free from genetically modified [ingredient]' or 'contains no genetically modified [ingredient],' be used" (Institute of Grocery Distribution 1997).
15. *The Independent on Sunday*, 21 March 1999.

References

Callon, Michel (1998) *The Laws of the Markets*. London: Blackwell Publishers. 26–7.

genetiX snowball (1998) *Handbook for Action*. One World Centre, 6 Mount Street, Manchester, M2 5NS.

Grove-White, Robin, Phil Macnaghten, Sue Mayer and Brian Wynne (1997) *Uncertain World: Genetically Modified Organisms, Food and Public Attitudes in Britain*. Lancaster: Centre for the Study of Environmental Change.

Harvey, Mark (1999) "Genetic Modification as a Bio-Socio-Economic Process: One Case of Tomato Purée," *Manchester*: CRIC.

Holliman, Jonathan (1974) Consumers' Guide to the Protection of the Environment Friends of the Earth. London: Cox and Wyman.

Hornig Priest, Susana (2000) *Grain of truth: the media, the public, and biotechnology*. Ithaca: Cornell University Press.

Institute of Grocery Distribution (1997) *Communication and labelling Guidelines for Genetically Modified Foods. Conclusions of a Four Year Research Programme*. Letchmore Heath, Watford, United Kingdom.

Marchand, Roland (1998) "Customer Research as Public Relations: General Motors in the 1930s," in *Getting and Spending: European and American consumer societies in the twentieth century*, Susan Strasser, Charles McGovern, and Matthias Judt (eds.). Cambridge: Cambridge University Press, 85–109.

Martineau, Belinda (2001) *First Fruit: The Creation of the Flavr Savr Tomato and the Birth of Biotech Foods*. New York: McGraw-Hill.

Miller, Peter, and Nikolas Rose (1997) "Mobilizing the Consumer: Assembling the Subject of Consumption," *Theory, Culture & Society*, 14(1): 1–36.

Robinson, Mike (1992) *The Greening of British Party Politics*. Manchester: Manchester University Press.

Sadler, Michèle (2000) *GAT Foods: Past, Present, Future? Industry's Approach. Consumer Attitudes. Expectations for the future*. Watford, United Kingdom: Institute of Grocery Distribution.

Williams, Raymond ([1983 [1976]) *Keywords: A vocabulary of culture and society*. New York: Oxford University Press, 78–9.

Part IV

Issues in Knowledge Politics as a New Political Field

Introduction

Nico Stehr

The struggle over the governance of knowledge in modern societies takes place within the context of very different social institutions. It takes place within professional life, the political system, the economy, social movements, special interest groups, the churches—but in the next decades increasingly also within transnational political bodies, non-governmental organizations, or multinational corporations.

The final section of our anthology deals with selected issues of knowledge politics as an emerging field of conflictual politics and policies in modern societies; in particular, it addresses some of the issues that are or are bound to arise in the legal system of advanced societies and the role experts, advisors and consultants are bound to play in the administrative system and in other institutions, and settings that have and will be forced to address and cope with the ways of regulating new knowledge.

Martin Schulte in his contribution deals with the use of knowledge in the legal system and the relation between scientific expertise and legal decisions. His focus is on the experience of the German legal system. Set within the theoretical framework of Niklas Luhmann's sociological theory of law, Schulte points out that references to and consultation of scientific experts have become a common, even routine and certainly indispensable feature of the complex administrative law of risk management, especially in the field of environmental law and legal norms pertaining to technological and scientific development in biotechnology, pharmaceuticals, etc.

Taking cues from Luhmann's theory of the legal system, Schulte aims for an "external description of the legal system as a self-describing system" (for example, in the form of legal practice, legal philosophy, legal dogmatics, and legal theory). Depending on the perspective chosen, the legal system is reconstructed in different ways. The approaches of the different sub-systems such as legal dogmatics are described. This approach then is employed or projected onto the ways in which the legal system uses knowledge, expertise, lack of knowledge and uncertain knowledge. What is involved in the case of administrative law amounts to a "dematerialization of law" demonstrated, for example, by the tendency of the legislature to postpone or abstain from risky decisions: more precisely, a dematerialization of legal norms invokes standards found outside of the boundaries of the legal system, such as "the state of art of scientific knowledge" or "recognized findings of scientific research." Such ambivalent and often highly contested legal norms with respect to the state of scientific knowledge have given rise to an almost routine consultation of experts by actors in the legal system. Although the manifest function of any consultation of experts is to reduce uncertainty, it often results in an increase of uncertainty or, as Luhmann observes, the experts are the "expressway to mutual irritability." The loose coupling of the legal system, the political system and the scientific community therefore represents one of the most difficult challenges of the governance of knowledge in modern, highly differentiated societies.

Lawrence Lessig takes up the theme of the relations between information, knowledge and the legal system when he focuses on the evolution of the Internet and the "innovation commons." Lessig's thesis is that the extraordinary innovations brought about by and for the Internet are based, last but not least, on the presence of core resources of the Internet on the "commons" of the Internet. The danger now is, he warns, that assertions of ownership of some of the same resources will bring about the demise of the innovative capacity of the Internet. The governance of the commons built into the very structure of the Internet threatens the diversity of creativity and multiplicity of opportunities the Internet has represented to date. With the ascendancy of legal regulation and state control, the Internet loses its status as a public good.

According to Lessig, the bias found in many cultures against common goods makes good sense in the case of those resources that are quickly depleted if left on the "commons." In the case of other resources, the "tragedy of the commons does not apply. Knowledge and information would be such a resource. For example, the wide dissemination of Max Weber's theory of the origins of capitalism does not deplete if, as a matter of fact, its wide dissemination is desirable.

Accordingly, the development of the Internet benefited from such the principle of easy access to its core resources. The control of the development of the Internet was not centralized or based on a top down but bottom up philosophy. The idea was to allow for flexibility and the consequence was innovation that originated at the end of the line.

However, not all features of the Internet are public goods. It is the "code layer" of the Internet sitting on top of a proprietary physical structure that was organized as a commons. The code layer does not have a bias toward specific contents, novel application programs, or other ways of using the Internet. As Lessig therefore puts it, "open code thus builds a commons for innovation at the content layer."

Developments, especially on the political front, now point in the direction of increasing governance of those features of the Internet that used to be on the commons; this applies both to the delivery of contents and the technological ways in which the contents are distributed. By moving central features of the Internet away from being common goods, the conditions for the possibility of innovation are affected. The short history of the lack of regulation and the growing regulatory environment of the Internet demonstrates the close interaction between the regulation of the new knowledge and the production possibilities for the subsequent production of novel knowledge in societies.

In the third contribution to the section. Stephen Turner examines the issue of governance of scientific knowledge in a comparative perspective. For practical purposes, what exactly is, or can be, the role of the public in the field of knowledge politics? In democratic societies there can be no doubt that the role of the public in either pressing for or sanctioning knowledge policies is of considerable importance. The reference to the role of the public is also a useful reminder that the kinds of practical choices knowledge politics may represent—as it is both drafted and implemented—are embedded not only in dynamic political and economic settings, but also in perhaps somewhat less dynamic sociocultural settings. As a result, knowledge policies are conditioned by a complicated host of factors, of which the notion of the public is one of the useful metaphors.

Observations about the relations of the public to science range from the straightforwardly empirical to contested normative or political analyses of the role of the public in scientific matters to dismissive or discouraged comments on the ability or inability of the public to engage with modern science. But the governance of scientific knowledge is not only linked to concerns and demands about the role of the public but accountability of science, the involvement of experts, bureaucratic institutions, and political cultures. The practical difficulty faced in this domain is that competent and efficient decision-making about the scientific knowledge requires such knowledge but that scientific knowledge is not sufficient to make such decisions. But can science advance such decisions?

Turner offers the argument that differences m national administrative traditions in particular have led to different models of the governance of knowledge in liberal democracies. The model of the governance of knowledge Turner discusses in particular relies on patterns of delegation of the tasks to specialists. He exemplifies the issues at hand with reference to the cholera epidemic in late nineteenth century Germany and shows how the influence of bureaucratic politics played a significant role in following or rejecting the advice of experts in different social and political settings. In short, the emerging combination of politics, expertise, and bureaucratic structures is decisive for knowledge politics.

As these contributions indicate too, the political landscape is changing as a result of new scientific discoveries and new technological innovations. The kind of regulative knowledge politics now in demand is new. Present mechanisms and institutions are unprepared to cope. But governments will be forced to face up to new problems and novel standards: they will have to develop new rules and they will be judged as to whether they are successful in meeting new goals. The nation-state will continue to be of consequence, but less so as an autonomous corporate actor that shapes knowledge politics. The knowledge politics of the nation-state will frequently have to, or (in the case of deliberate policy transfers), desire to enact policies of wider global institutions, international treaties and social movements. However—and this can also already be detected—the tempo with which solutions to the new problems are found will be far outpaced by the accumulation of new political challenges.

10

The Use of Knowledge in the Legal System: The Relationship between Scientific Expertise and Legal Decisions

Martin Schulte

Since the 1970s at least, the consultation of scientific expertise has been a common, mostly indispensable part of complex administrative procedures, particularly in the field of administrative law of risk management, such as environmental law and law of technology, biotechnology, drug manufacturing, etc. (Di Fabio 1990). Using vague legal terms, such as "state of science and technology," it is established that the legal system is "dependent" on the progress of research (Weingart 2001). In the following, the use of knowledge in the field of law, as expressed in the relationship between scientific expertise and legal decision, shall be analyzed on the basis of a sociological theory of law, by giving an external description of the legal system as a self-describing system.

Fundamental Methodological Assumptions of a Sociological Theory of Law

The approach upon which these considerations are based methodologically refers to a stimulus given by Niklas Luhmann in his "Law of the Society": "Accordingly, a sociological legal theory would come down to an external description of the legal system, however, it would represent a theory which is appropriate to the matter only in case it described the system as a self-describing system (which has hardly ever been tried out in today's sociological jurisprudence)" (Luhmann 1993: 17).

Thus, we are aiming a sociological theory of law in the sense of an external description of the legal system as a self-describing system. This requires clarification on the operations and different perspectives of self-description and external description of the legal system (legal practice, legal dogmatics, legal methodology, legal philosophy, legal theory; Schulte 1993: 317–32).

But what do the operations of self-observation/description and external observation/description of the legal system depict? First of all, they are operations of communication, which exist only within the context of events in the system. They provide a double description of each sub-system in society, including the legal system: a self-description (description from the inside, in particular by legal dogmatics and legal philosophy) on the one hand, and an external description (from the outside, by a sociological theory of law) on the other. Reference is always made to one and the same system. The only difference is in the inclusion of the description in, or its exclusion from, the described system. There may be mutual irritability between the internal and external descriptions of the legal system, since communication beyond system limits existing in the society remains possible in the life of society (Kieserling 2000: 38, 39; Luhmann 1993: 496 f.; Luhmann 1997: 883).

Self-observation and self-description of the legal system are of the utmost importance. Both of them have as their aim the unity of the system. Self-observation is the operation directed towards the system and carried out within the system, while self-description refers to the production of semantic artifacts. In the course of their implementation in the actual communication, they inevitably become a subject of observation and description themselves. The problem of self-observation and self-description is that the "problem of identity" is preserved in the identity of the problem. However, it must not be overlooked that each solution to a problem, each identity proposal, is made as an operation of the system and therefore inevitably becomes a subject of observation within the system (Luhmann 1997: 879 ff).

Based on our methodological principle explained above, which aims at an external description of the legal system as a self-describing system, this self-description must be given particular attention: we must define the terms more precisely and outline their contents. Above all, self-descriptions are observations in the strict sense of the word. Each self-description of the system is always a construction by which the system deals with perceived inconsistencies, thus limiting and increasing irritability at the same time. Furthermore, there may be numerous conceptual self-descriptions of the system. The example of the legal system illustrates this very well, since legal dogmatics and legal philosophy, possibly even legal practice, produce different self-descriptions of the same system. Each of these self-descriptions takes its own contingency into consideration, and also the fact that there may be other self-descriptions of the system (Luhmann 1997: 882, 886, 891 f.).

As mentioned above, the aim of self-description is the "representation of the unity of the system within the system." It is the reflection of unity within the system that reflects itself. For this reason, it must meet the system's requirements, show regard for it and accept its characteristic features. Therefore, self-description of the legal system must comply with its own standards. It is forced to allocate itself to the system it describes by accepting and stressing the system-specific connections. Consequently, it is impossible for legal dogmatics to deny that acting in conformity with legal rules is right. The basic distinction between norm and factum is of great importance to the legal system, particularly in regard to normativity (Luhmann 1993: 498, 501 f.).

Against this background one can now deal with the different perspectives of observation and description, that is, legal practice, legal dogmatics, legal methodology, legal philosophy, and legal theory, all of which deal with the legal system, however, at different levels of abstraction and from their specific perspective. Depending on the perspective, the legal system is reconstructed in different ways (Krawietz 1992: 27 f.).

In the external description of a sociological theory of law, *legal practice* is the organization (Luhmann 2000) of judicial and extra-judicial sectors of the legal system (e.g., jurisdiction, legislation), but also of constitutive interaction (Kieserling 1999), such as conclusion of contracts, within the legal system. It works on the foundation of basal self-reference, the distinction between element and relation (Luhmann 1984: 600 f.). The self-referring itself is legal communication in law (e.g., in the form of a court judgment, the filing of a petition to a court, or the declaration of the intention to conclude a contract).

By organizing the judicial sectors of the legal system, a closer sector of legally binding decisions is created as a condition of continuous observation of observing, an organized sub-system broken down even further by the distinction between members and non-members. Its members, judges, but also judicial officers outside the regular judiciary, continuously produce decisions based on a universal code of legal/illegal and following the standards of decision-making programs (constitution, law, ordinance, etc.), so that there is good reason to designate this as an "organized decision-making system of the legal system." It functions on the basis of reflexivity by double modalization, that is, it is a matter of standardizing standardization, which becomes particularly clear in situations when decisions of the (German) Federal Constitutional Court are given the force of binding law in the cases mentioned in Paragraph 31, Section 2. Sub-section 1 BVerfGG (Luhmann 1993: 144 ff.).

Legal dogmatics, on the other hand, reflects legal practice as a self-abstraction. In the sense of a "consistency control," it defines the "conditions of what is juridically possible," that is, the possibilities of the juridical construction of legal matters (Luhmann 1974: 18 f.). It works on the basis of processual self- reference (reflexivity), using the distinction of "before" and "after." In this case, the self-referring is not an aspect of the distinction, but a process constituted by it. Communication corresponds to this, because its elemental events are determined by the expectation of a reaction and reaction to an expectation (Luhmann 1984: 601). In particular, this applies to legal communication in law. Law "operates reflexively," the "basic distinction between cognitive and normative expectation" itself is made the "sub-ject of normative expectation," or, to be brief, law "is only law; in case one can expect that normative expectation is normatively expected." Legal dogmatics regularly refer to *preceding* decisions of legal practice, or *preceding* statements of its own, so it communicates on the basis of (legal) communication. In doing so, legal dogmatics create normative draft versions of what is juridically possible under the existing real circumstances, that is, possibilities of the juridical construction of legal matters. In the sense of standardizing the standardization, it also addresses the organized decision-making system of the legal system, in particular the legislator, by making legally political suggestions on the further development of valid law.

Legal philosophy, finally, aims at the conceptual definition of the identity of the legal system in contrast to its environment, *inter alia* by making observations about the legitimation of the validity of positive law (legitimation by procedure, legitimation by giving reasons, legiti-mation by consensus, etc.). Since it is based on the distinction of system and environment, we can call this reflection (Kieserling 2000: 50 ff.). The self-referring *is* the system. However, in contrast to basal and pro-cessual self-reference, one can establish the exceptional case that even the conceptual domains of self-reference and system reference, that is, the operation describing a system by means of the distinction between system and environment, overlap each other (Luhmann 1984: 601).

Since legal philosophy develops a theory of the legal system, refers to the system and takes the same view, one can designate it very briefly as a "theory of the system within the system" (Luhmann 1999: 419, 422).

However, it would not be sufficient to confine oneself to this external description of the legal system from the perspective of a sociological theory of law. As explained before, the specific feature of our approach consists of an external description of the legal system *as a self-describing system*. Therefore, it is necessary to consider that the different sub-systems of the legal system, as an expression of their self-reference, in turn produce different self-descriptions reflecting their relationship to each other.

Succinctly, *legal practice* is concerned with the practical handling of law in everyday situations, in particular in the course of legal decision-making. The center of interest and attention is still the "judi-cial power." This does not only apply to the perception of jurisdiction in the public, but also to the orientation of training in the field of law. However, it must not be forgotten that legislation, application of law (government and administration), and, first of all, legal advice given by lawyers have equal claim to be considered constituent elements of legal practice.

Legal dogmatics, with slightly different nuances, sees itself as the "juridical research interest par excellence" or as the "core discipline of legal science" (Dreier 1971, 1990; Simon 1992). As such, it always emphasizes its special relation to legal practice; its research interest is determined by the performance "which legal practice, in particular judges, normally expect it to provide and what they reasonably may expect from it" (Dreier 1990: 21; Engel 1998; 26, 36; Morlok 1998: 7; Schünemann 1976: 23). Nevertheless, legal dogmatics is not firmly embedded in the general juridical language, nor does it always meet with a positive response. Instead, it is quite often associated with "conservative ossification," "dogmatism," and "remoteness of law from real life." Finally, there is no doubt that the term is still suffering from the burden of "adaptability" of legal dogmatics to changing political systems (Rüthers 1999: 176).

Today's perception of legal dogmatics can be traced back to papers written in the early 1970s by Josef Esser (1972: 90 ff.), Franz. Wieacker (1970: 311 ff.), Niklas Luhmann (1974: 15 ff.) and Winfried Brohm (1972). Although there are differences in terminological details, one can clearly notice common features in their basic ideas (Herberger 1981: 345 ff.; Herberger 1984: 91 ff.; Sandström 1998: 194; Laband 1911; ix; Dreier 1990: 22).

Dogmatics in general should be the "doctrine of established 'truth' put forward in an authoritative—in this case, legal manner, its connection to a system and its utilization for the recognition of law in the individual case" (Esser 1968: 95, 96; with a critical attitude Larenz 1991: 224 ff.). From time to time, it is also designated as the creation of a communicable system of terms, rules, principles, and institutions of positive law, "all of which taken together make up its set of dogmatic rules" (Gröschner et al. 2000: 2). This line is also followed by the perception of legal dogmatics as a "structure of legal terms, institutions, principles, and rules created within the system" that, as components of the positive legal system claim general recognition and obedience, whether or not laid down in law (Brohm 1972: 246; Aarnio 1979: 34; Aarnio 1988: 93, 94). Or it is stated:

> that those lawyers work dogmatically who aim at organizing the set of legal rules important for a certain field in accordance with uniform, comprehensive and consistent criteria and connections, in order to make terms and basic ideas of general significance found in the course of this work known, to organize them in a logical, and thus inter-subjectively communicable, system and to check the continuously produced legal matter—e.g., new legal provisions or court decisions—in order to establish whether it fits with the given order or requires this order to be changed or extended. (Kötz 1990 78; Zweigert 1969: 445)

Older papers in particular were based on the idea that *legal philosophy* represented reflections of legal values or the doctrine of legal values (Lask 1923: 286, 279 f.; Radbruch 1948: 17; Radbruch 1973: 93 ff., 96; Stammler 1928; Winkler 1969: 39). Also the more ontic definition of the term, considering legal philosophy to be the "science of the nature and idea of law, the cognition of law, the terms of law and the individual legal terms, furthermore, of the classification of legal sciences, of the aim and object of law and also of the legal world view" (Kubes 1982: 221), is now a matter of the past. However, in the case of language analysis (Krawietz 1996), which is still supported, legal philosophy is made the critical horizon of jurisdiction. Juridical value judgments existing in the language are seen against this background, discussed and systematized, in which legal philosophy is given the task to recognize the "right" law. The currently "prevailing" idea of legal philosophy clearly stands out against such definitions of terms characterized so firmly by a certain approach of legal philosophy. It very openly represents itself as the discipline reflecting and discussing the questions of fundamental juridical principles and general problems in a "philosophical manner." If possible, answers are given to these questions (Kaufmann 1989; Engel 1998: 26;

Simon 1992: 361; Weinberger 1988: 46) or law; what it should be like, is considered with normative intentions (Rüthers 1999: 8).

These are the contributions that can be made by a sociological theory of law; giving an external description of the legal system as a self-describing system. Now, this approach shall be projected onto the use of knowledge in law, which will be done on the example of the relationship between scientific expertise and legal decision in complex administrative procedures.

The Use of Knowledge in Law from the Perspective of a Sociological Theory of Law

Scientific Expertise and Legal Decision in the Self-Description of the Legal System—the Example of 'Electrosmog'

In the self-description of the legal system, the dogmatics of administrative law recently emphasized a fundamental trend in development characterizing the way in which law handles knowledge, lack of knowledge, and uncertain knowledge. This refers to the so-called "dematerialization of law," demonstrated by the legislator's tendency to "postpone risky decisions, to maintain its own scope of action wherever possible and to confine itself to preventing wrong decisions." Therefore, in order to put its purpose of protection in more concrete terms, dematerialized law frequently refers to the results of standardization outside the field of law (state of the art, state of science and technology, recognized findings of scientific research; Scherzberg 2002).

Paragraph 3, Section 6 of the German Law on ambient air (BImSchG) for example, defines the *state of the art* as the stage of development reached by progressive processes, facilities, or modes of operation that appear to be likely in practice to limit emissions. For the handling of substances contaminating water, however, it is laid down in Paragraph 19 g. Section 3 of the German Water Resources Law (WHG) that facilities must at least comply with the generally recognized standards of technology in terms of their design, installation, setup, maintenance and operation. As far as the transportation of nuclear fuels is concerned, it is established in Paragraph 4. Section 2, Sub-section 3 of the German Atomic Energy Law (AtG) that the permission must be granted if it is ensured that the nuclear fuels are transported in compliance with the legal rules governing the transportation of hazardous goods that are applicable to the relevant means of transport, or, in case such provisions do not exist, the permission must be granted if precautions against damage caused by the transportation of nuclear fuels have been taken in any other way in accordance with the *state of science and technology*. Finally. Paragraph 7, Section 2, Sentence 2 of the German Law on the Protection of Animals (TSchG) requires that any decision on the question whether animal experiments are indispensable must be based on the relevant *state of scientific knowledge*, and it must be checked if the intended aim could not be achieved using other methods or processes. And one can also find regulations concerning the field of "electrosmog," which is of particular interest to our questions here. Paragraph 5, Sentence 1 of the 26th Ordinance on the Law of ambient air (BImSchV) requires that measuring instruments, measuring and calculation methods used in the determination of the electric and magnetic field strength and the magnetic flux density, including the consideration of existing emissions, must comply with the "state of measuring and calculation techniques." In Paragraph 2, Number 9 of the draft version of an Environmental Code, the legislator even attempted to give a general definition not restricted to a particular field, saying that the state of technology refers to the stage of development reached by effective and progressive processes, facilities or modes of operation that appear to be likely in practice to spare natural assets and other resources or to prevent or, if this is impossible, minimize any adverse impacts on the assets protected by this code.

Experts are often consulted in administrative procedures and judicial proceedings to fulfill such "vague terms." They are generally expected to "provide the scientific background with respect to certain facts of the case," "to ascertain in an expert manner the facts necessary for the legal assessment of the circumstances and to inform about them" and "to assess and evaluate certain circumstances by means of their expert knowledge." However, the determination and interpretation of the applicable law as well as the ascertainment of the merits of the case are left up to the authorities and the courts. Such tasks must not be assigned to the expert. The same applies to the question, if the permission to build, or to operate, technical facilities is in accordance with the "state of science and state of the art" as required by the applicable law, and if optimal protection against dangers and precautions against risks are ensured in this way. All this is derived from the principle of due course of law and the fact that administration and jurisdiction are bound by law and justice (Section 20, Subsection 3 of the German Basic Law; Breuer 2001: 43 f.).

Insofar as the authority, fulfilling the "vague terms," has accepted an expert's evaluation given in his expertise as its own assessment and has made the expert opinion the basis for its decisions, the question of judicial review of the administration's decision is raised, which is still disputed in the dogmatics of constitutional and administrative law. In this dispute, even in case one takes up a position against granting the administration a generous scope for judgment evaluation (Schröder 2002: 185) and in favor of unlimited judicial review; it is only to be found out if the reasons that induced administration to take its decision are based on reliable facts, are free from inconsistencies, and plausible (Kutscheidt 2001: 103).

The legal practice at the highest law courts is even more restrictive in this respect. With regard to the individual rights of protection against possible risks for health caused by "electrosmog," the German Federal Constitutional Court just recently pointed out that the authorities issuing ordinances "have a wide scope of assessment, evaluation and discretion." In such a "situation of uncertainty," it is solely within the "political discretion" of such authority if it wishes to take precautions, so to speak, "at random." Even the obligation of the state to protect its citizens does not justify the enforcement of uncertain scientific findings by means of processual law, it does not give any good reason to keep the decisions of authorities issuing ordinances on precautions under control nor to assess whether limit values provide sufficient protection, based on the latest findings of research. Instead, it is up to the authorities issuing ordinances to follow the progress of science in any respect using suitable means, and to evaluate it so that further protective measures can be taken if necessary. According to the German Federal Constitutional Court, this accounts for the different possibilities in function and procedure of administration and jurisdiction to gain knowledge for the evaluation of complex situations with dangerous potential subject to disputes among scientists. It is emphasized that the research on the effects of electromagnetic fields on human beings, carried out for some time on an international and interdisciplinary basis, is by no means complete, so that the consideration of individual scientific studies does not provide any consistent picture of the dangerous potential involved. The interdisciplinary examination and evaluation of the wide range of research work is in the hands of different international and national professional committees, such as an expert group at the German Federal Radiation Protection Office. Taking this background into consideration, it is impossible to "comprehensively assess the state of complex scientific expertise as required" when evidence is taken at court on the occasion of one particular dispute. The courts will only be able to assess risks independently "when research has reached a level which allows to confine the problems to be evaluated to certain questions which can be clarified by acknowledged scientists on the basis of recognized findings" (Deutsches Verwaltungsblatt [DVB; German Federal Constitutional Court] 2002: vol. 117, pp. 614, 615).

Judicial control of complex administrative decisions is also complicated by the "merit of practical reason," which is to be applied in the evaluation of technical dangers and risks. On the one hand, this refers to "gaming practicable and justifiable legal principles and maxims of action in view of technical dangers and risks as well as uncertainties immanent in prognoses given by natural sciences and engineering due to their nature"—which is to be understood purely pragmatically and not epistemologically. The "fundamental principles and limits to the required minimization of risks as laid down in standards, legal rules and the constitution as well as the residual risks to be tolerated are to be described so that legal practice can become aware of them and perform accordingly" (Breuer 2001: 56 f.). One objection against this view is that it is inevitable to determine the "merit of practical reason" as an epistemological pattern of thought and to derive legal consequences for the activities of experts from it. Therefore, the objective to be achieved in collective decisions concerning dangers and risks must be the determination of the "most superior justification of law on the basis of reason or other moral motivation." Consequently, the "merit of practical reason," if connected with the "claim for correctness of collective legal decisions as regards content" requires an "idealistic basic standard of reasonable explanation"; the reason why such a standard could create an obligation can only be "the connection of idealistically categoric and utilitarian purposeful reasoning" (Bartlsperger 2000: 149 ff.; Bartlsperger 2001: 121 ff.). However, it is worth stating that the overwhelming majority of judges in the field of administrative law have not followed this line of interpreting legal rules following the ideas of legal philosophy, but have adopted the pragmatic approach to interpretation instead (DVB1: 49, 89, 134 ff.; Marburger 1985: 95 ff.; Nolte 1984: 62 ff.; Sommer 1981; Winter and Schäfer 1985).

Scientific Expertise and Legal Decision in the External Description of the Legal System

Before we focus the external description of the legal system on the example of scientific expertise and juridical decision, it should be mentioned that the fulfilling of "vague terms," for example, the term of the "state of science and technology," by the administration represents a relatively new manifestation of a structural coupling (Luhmann 1997: 92 ff., 776 ff.; Schemann 1992; Lieckweg 2001) of the scientific and political system. This refers to experts providing consultative service to politicians (Luhmann 1997: 785 f.), which first came into existence in connection with the risk management developing in the 1970s and is presently gaining particular importance and dynamism with respect to the whole field of biotechnology, but also to the relatively new phenomenon of "electrosmog" as outlined above.

Consultative service granted to politicians by experts no longer means that they simply have to apply existing knowledge. Nowadays, they have to master safety problems the complexity of which overtaxes administration and jurisdiction. Therefore, experts should serve the absorption of uncertainty, but paradoxically enough, it is their consultance that again produces uncertainty (Luhmann 1994: 32 f.; Luhmann 2000a: 161 ff.; Weingart 2001: 157 ff.; Nowotny, Scott, and Gibbons 2001: 215 ff.; Jasanoff 1995). According to Niklas Luhmann, one might therefore designate them as an "expressway to mutual irritability" (1997: 786).

What manifests itself as a structural coupling in the relationship of science and politics must be analyzed in a more specific way with regard to the relationship of the scientific and legal system. If the interpretation of "vague terms," for example, the term of the "state of science and technology," requires the consulting and application of scientific expertise in the course of legal proceedings, these normally include the judges, the persons taking part in the proceedings (parties to the lawsuit and their legal advisers) and the expert. As seen from the perspective of the external description of the legal system, communication between the persons present takes place on the basis of its programs (e.g., the German Law on ambient air), following the code of legal/illegal. One can therefore designate this as interaction within the organized system of decision-making of the legal system (Luhmann 1997: 812 ff., 814; Luhmann 1993: 145; Kieserling 1999: 15 ff.). Although jurisdiction, as part of the functional system of law, has its own unambiguous identity, it should be mentioned that in the event that scientific expertise is consulted and integrated into legal proceedings, organizational communication does not only involve the legal system, but also the system of science. This is partly understood as the "multi-referential" capability of organizations (Wehrsig and Tacke 1992), while Luhmann interprets this as a manifestation of "loose coupling" providing a "meeting place for the different functional systems" (2000a: 398; also Lieckweg 2001: 273 f.).

By the variety of "vague terms," the legal system has an open cognitive attitude towards its environment. When scientific expertise is consulted, it is irritated right here by scientific, knowledge, which we should not assume to be a "stock (of knowledge) invariable in the course of time" (Luhmann 1990: 129), which can be transferred from one system to another like an object. Although knowledge appears in an objectified form in order to be lasting, it is always a current operation, as in the construction of generalized cognitive expectations (Stichweh 2000: 132), which, however, in the course of being carried out disappears again. Knowledge can become time lasting only through the performance of an observer (e.g., the legal system), whose operation of observing, however, is again subject to the general rules of any sort of operating (Luhmann 1990: 129, 130).

The interpretation and application of 'vague terms' turns out to be an operative coupling of the legal system with the system of science. Here, the term of "operative coupling" is to be understood as the counterpart of "structural coupling." While the latter means that "a system always presumes certain features of its environment to be time lasting and relies on them in its structure," the former refers to the "coupling of operations by operations." One of the two versions of operative coupling is "the simultaneity of system and environment which is always to be presumed." It facilitates the momentary coupling of operations of the system with those that are allocated to the environment by the system (Luhmann 1993: 440 f.). In the course of a judge's examination of scientific expertise with regard to inconsistencies and plausibility, the determination of the merits of the case, the interpretation of the applicable law and the concrete subsumption under the "vague term," the system of science is observed by the legal system for the duration of a single event, that is, an operative coupling of these two functional systems is established.

But right here, the cognitive openness correlating with the operative closure of the legal system manifests itself, since the system of science is observed by the legal system on the basis of its own code and its own programs. This is particularly clearly demonstrated by the above mentioned decision of the German Federal Constitutional Court concerning the dangers of the so-called electrosmog, where the court answers the question whether the authority issuing ordinances must tighten the currently applicable emission limits to protect the population against hypothetical dangers, and under what circumstances courts are obliged to provide evidence for the claim that the valid limit values are obsolete in view of new scientific findings about the danger of emissions. Regarding this matter, express reference is made to the theories of discretionary powers, the obligation of the state to protect its members, and the obligation of authorities issuing ordinances to improve the same, which have been developed in legal practice, in particular by the jurisdiction of the highest law courts, such as the German Federal Administrative Court and the German Federal Constitutional Court, and also by legal dogmatics. Thus, legal practice with its operations (observations) takes up its own operations (observations); and it is this autopoiesis in which the second version of operative coupling is implemented (Luhmann 1993: 440).

The autopoiesis of the specifically coded functional systems (legal system and system of science) is also the reason for the differences in the 'level of demands' on the use of knowledge. The system of science, making the concession of remaining uncertainty, on the one hand expresses the basic structure of cognitive expectation, that is, expectation has to be corrected in the event it fails to come true (Luhmann 1990: 138), on the other, it aims at protecting the results of research against criticism, just like an "additional value," while the legal system, primarily strives to maintain its owti decisions. As far as the certainty, or uncertainty, of scientific knowledge is concerned, the legal system can therefore waive the description of the fundamental uncertainty of all knowledge more easily. But it formulates higher demands on the certainty of scientific knowledge if punishable offences, liability (Hapke and Japp 2001) or damages are concerned in the field of law (Luhmann 1993: 91).

Finally, it is worth stating that the use of (scientific) knowledge challenges the legal system in a very specific way. But thanks to its cognitive openness and operative closure, it proves that it is able to stand up to this challenge (Schulte 2002, 2003). In the end, everything comes down to the question of (internal/external) observation: Observing changes the world in which people observe (Luhmann 1990: 75).

References

Aarnio, Aulis (1979) *Denkweisen derRechtswissenschaft.* Vienna: Springer-Verlag.
——— (1988) "Rechtsdogmatik: Wisscnschaft oder Technik?" in *Wegen Recht und Billigkeit: Vorträge und Aufsätze aus 10 Jahren,* Aulis Aarnio. Berlin: Duncker and Humblot. 90–102.
Bartlsperger, Richard (2000) "Ökologischc Gewichtungs- und Vorrangregelungen," in *Planung, Festschrift für Werner Hoppe zum 70. Geburtstag,* Wilfried

Erbguth, Janbernd Oebbecke, Hans-Werner Rengeling, and Martin Schulte (eds.). Munich: C.H. Beck. 127–52.

—— (2001) "Das Vemunftmotiv der Ungebundenheit im Expertenwesen," in *Der Experte bei der Beurteilung von Gefahren und Risiken: Vorträge auf der gleichnamigen Veranstaltung vom 17./18. November 1995 an der Universitat Erlangen-Nümberg in Erlangen*. Richard Bartlsperger (ed). Berlin: Duncker and Humblot. 117–27.

Breuer, Rüdiger (2001) "Die Angst vor Gefahren und Risiken und die sachverständige Beratung nach dem Maßstab praktischer Vernunft," in *Der Experte bei der Beurteilung von Gefahren und Risiken: Vorträge auf der gleichnamigen Veranstaltung vom 17./18. November 1995 an der Universität Erlangen-Nümberg in Erlangen.*, Richard Bartlsperger (ed.) Berlin: Duncker and Humblot. 31–79.

Brohm, Winfried (1972) "Die Dogmatik des Verwaltungsrechts vor den Gegenwartsaufgabcn der Verwaltung." *Veröffentlichungen der Vereinigung der deutschen Staatsrechtslehrer* 30: 245–306.

Di Fabio, Udo (1990) "Verwaltungsentscheidung durch extemen Sachverstand," *Verwaltungsarchiv* 81:193–227.

Dreier, Ralf (1971) "Zum Selbstverstandnis der Jurisprudenz als Wissenschaft," *Rechtstheorie* 2: 37–54.

—— (1990) "Rechtstheorie und Rechtsgeschichte," in *Rechtsdogmatik und praktische Vemunft*, Okko Behrends, Malte Dießelhond, and Ralf Dreier (eds.). Gottingen: Vandenhoeck and Ruprecht. 17–34.

Engel, Christoph (1998) "Rechtswissenschaft als angewandte Sazialwissenschaft," in *Gemeinschaftsgüter: Recht, Politik und Okonomie*, Preprints aus der Max-Planck-Projektgruppe Recht der Gemeinschaftsgüter, Bonn.

Esser, Josef (1968) "Das Bewussrwerden wissenschaftlichen Arbeitens im Recht," in *Dubischar, Grundbegriffe des Rechts*. Stuttgart: Kohlhammer-Verlag. 95–102.

—— (1972) *Vorverstdndnis und Methodenwahl in der Rechtsfindung*. 2nd Edition. Frankfurt am Main: Fischer Taschenbuch Verlag.

Gröschner, Rolf, Klaus Dierksmeier, Michael Henkel, and Alexander Wiehart (2000) *Rechts- und Staatsphilosophie. Ein dogmenphilosophischer Dialog*. Berlin: Springer-Verlag.

Hapke, Uwe, and Klaus Japp (2001) *Prävention und Umwelthaftung. Zur Soziologie einer modernen Haftungsform*. Wiesbaden: Deutscher Universitätsverlag.

Herbexger, Maximilian (1981) *Dogmatik, Zur Geschichte von Begriff und Methode in Medizin und Jurisprudenz*. Frankfurt am Main: Vittorio Klostermann.

—— (1984) "Logik und Dogmatik bet Paul La band. Zur Praxis der sog, jurtstischen Methode im 'Staatsrecht des Deutschen Reiches'," in *Missenschaft und Recht der Verwaltung seit dem Ancien Régime*, Erk Volkmar Heyen (ed.). Frankfurt am Main: Vittorio Klostermann. 91–104.

Jasanoff, Sheila (1995) *Science at the Bar Law, Science and Technology in America*. Cambridge. Harvard University Press.

Kaufmann, Arthur (1989) "Rechtsphilosophic, Rechtstheorie, Rechtsdogmatik," in *Einführung in Rechtsphilosophie und Rechtstheorie der Gegemnart*,

Arthur Kaufmann, Winifried Hassemer (eds.). 5th edition. Heidelberg: UTB C.F. Müller. 1–24.

Kieserling, Andre (1999) *Kommunikation unter Anwesenden: Studten über Interaktions systeme.* Frankfurt am Main; Suhrkamp.

—— (2000) "Die Soziologie der Selbstbeschreibung," in *Rezeptiem und Reflexion: Zur Resonanz der Systemtheorie Niklas Luhmanns außerhalb der Soziologie,* Henk de Berg and Johannes Schmidt (eds.). Frankfurt am Main: Suhrkamp. 38–92.

Kötz, Hein (1990) "Rechtsvergleichung und Rechtsdogmatik," in *Rechtsdogmatik und Rechtspolitik: Hamburger Ringvorlesung.* Karsten Schmidt (ed.). Berlin: Duncker and Humblot. 75–89.

Krawictz, Werner (1992) "Zur Einführung: Neue Sequenzierung der Theoriebildung und Kritik der allgemeinen Theorie sozialer Systeme," in *Kritik der Theorie sozialer Systeme,* Werner Krawietz and Michael Wielker (eds.). Frankfurt am Main: Suhrkamp. 14–42.

—— (1996) "Sprachphilosophie in der Jurisprudenz," in *Sprachphilosophie. Ein internationales Handbuch zeitgenössischer Forschung,* Volume II. Marcelo Dascal, Diettned Gerhardus, Kuno Lorenz, and Georg Meggle (eds.). Berlin: de Gruyter. 1470–89.

Kubes, Wladunir (1982) "Rechtstheorie und Rechtsphilosophie," *Rechtstheorie* 13: 207–25.

Kutscheidt, Ernst (2001) "Bewertung von Sachverständigengutachten in Verwaltungs- und Genchtsverfahren," in *Der Experte bei der Beurteilung von Gefahren und Risiken,* Richard Bartlsperger (ed.). Berlin: Duncker and Humblot. 93–104.

Laband, Paul (1911) *Das Staatsrecht des Deutschen Reiches.* 5th Edition, Volume I. Tübingen: Mohr Siebeck Verlag.

Larenz, Karl (1991) *Methodenlehre der Rechtswissenschaft.* 6th Edition. Berlin. Springer-Verlag.

Lask, Emil (1923) *Gesammelte Schriften.* Eugcn Herrigel (ed.). Volume I. Tübingen: Mohr Siebeck.

Lieckweg, Tania (2001) "Struklurelle Kopplung von Funktionssystemen 'über' Organisation." *Saziale Systeme* 7: 267–89.

Luhmann, Niklas (1974) *Rechtssystem und Rechtsdogmatik.* Stuttgart: Kohlhammer Verlag.

—— (1984) *Soziale Systeme. Grundriss einer allgemeinen Iheorie.* Frankfurt am Main: Suhrkamp.

—— (1990) *Die Wissenschaft der Gesellschaft.* Frankfurt am Main: Suhrkamp.

—— (1993) *Das Recht der Gesellschaft.* Frankfurt am Main: Suhrkamp.

—— (1994) "Wessen Umwelt?" in *Wissenschaften im ökologischen Wandel,* Umweltbundesamt *[Federal Environment Agency]* (ed.). Berlin: Selbstverlag. 25–33.

—— (1997) *Die Gesellschaft der Gesellschaft.* Frankfurt am Main: Suhrkamp.

—— (1999) *Ausdifferenzierung des Rechts. Beiträge zur Rechtssoziologie und Rechtstheorie.* Frankfurt am Main: Suhrkamp.

—— (2000a) *Die Politik der Gesellschaft.* Frankfurt am Main: Suhrkamp.

—— (2000b) *Organisation und Entscheidung.* Wiesbaden Westdeutscher Verlag.

Marburger, Peter (1985) *Atomrechtliche Schadensvorsorge.* 2nd Edition. Cologne: Carl Heymanns Verlag.

Morlok, Martin (1998) "Vom Reiz und vom Nutzen, von den Schwierigkeiten und den Gefahren der Ökonomischen Theorie für das Öftentlichc Recht," in *Öffentliches Recht als ein Gegenstand öktwtomischer Forschung,* Christoph Engel and Martin Morlok (eds.). Tübingen: Mohr Siebeck Verlag. 1–29.

Nolte, Rüdiger (1984) Rechtliche Anfonderung an die technische Sicherheit von Kernanlagen: Zur Konkretisierung des § 7 Abs. 2 Nr. 3 AtomG. *Berlin: Duncker and Humblot.*

Nowotny, Helga, Peter Scott, and Michael Gibbons (2001) *Re-Thinking Science, Knowledge and the Public in an Age of Uncertainty.* Cambridge: Polity Press.

Radbruch, Gustav (1948) *Vorschule der Rechtsphilosophie* Heidelberg: Scherer.

—— (1973) *Rechtsphilosophie,* Erik von Wolf and Hans-Peter Schneider (eds.). 8th Edition. Stuttgart: Koehler.

Rüthers, Bernd (1999) *Rechtstheorie, Begriff, Geltung und Anwendung des Rechts. Munich: C. H. Beck.*

Sandström, Marie (1998) "Das dogmatische Verfahren als Muster der rechtswissensehaftlichen Argumentation," in *Entwicklung der Methodenlehre in Rechtswissenschaft und Philosophie vom 16.–18. Jahrhundert,* Rainer Schröder (ed.). Stuttgart: Steiner Verlag. 191–203.

Schemann, Andreas (1992) "Strukturelle Kopplung. Zur Festlegung und normativen Bindung offener Möglichkeiten sozialen Handelns," in *Kritik der Theorie sozialer Systeme,* Werner Krawietz and Michael Welker (eds.). Frankfurt am Main: Suhrkamp. 215–29.

Scherzberg, Arno (2002) "Wissen, Nichtwissen und Ungewissheit im Recht," in *Wissen, Nichtwissen, Unsicheres Wissen,* Christoph Engel, Jost Halfmann, and Martin Schulte (eds.). Baden-Baden: Nomos Verlag. 113–44.

Schröder, Rainer (2002) "Verfassungsrechtliche Grundlagen des Technikrechts," in *Handbuch des Technikrechts.* Martin Schulte (ed.). Berlin: Springer-Verlag. 185–208.

Schulte, Martin (1993) "Recht, Slaat und Gesellschaft—rechlsrealistisch betrachtet," in *Rechtsnorm und Rechtswirklichkeit. Festschrift für Werner Krawietz zum 60. Geburtstag,* Aulius Aarnio, Stanley L. Paulson, Ota Weinberger, Georg Henrik von Wright, and Dieter Wyduckel (eds.). Berlin: Duncker and Humblot. 317–32.

—— (2002) "Zum Umgang mit Wissen, Nichtwissen und Unsicherem Wissen im Recht — dargestellt am Beispiel des BSE- und MKS-Konflikts –," in *Wissen, Nichtwissen, Unsicheres Wissen,* Christoph Engel. Jost Halfmann, and Martin Schulte (eds.). Baden-Baden: Nomos. 351–70.

—— (2003) "Wissensgenerierung in Recht und Rechtswissenschaft — Selbstbeschreibung und Fremdbeschreibung des Rechtssytcms –," in *Szenarien der Wissensgesellschaft,* Armin Nasschi (ed.). Frankfurt am Main: Suhrkamp. In press.

Schünemann, Hans-Wilhelm (1976). *Sozialwissenschaften und Jurisprudenz.* Munich: C. H. Beck Verlag.

Simon, Dieter (1992) "Die Rechtswissenschaft als Geisteswissenschaft," *Rechtshis-torisches Journal* 12: 351–66.

Sommer, Herbert (1981) "Praktische Vernunft beim kritischen Reaktor," *Die Öffentliche Verwaltung:* 34: 654–60.

Stammler, Rudolf (1928) *Lehrbuch der Rechtsphilosophie.* 3rd Edition. Berlin, Leipzig: Walter de Gruyter.

Stichweh, Rudolf (2000) "Kultur, Wissen und die Theorien soziokultureller Evolution," in *Ortsbestimmungen der Soziologie: Wie die kommende Generation Gesellschaftswissenschaften betreiben will,* Ulrich Beck and Andre Kieserling (eds.). Baden-Baden: Nomos Verlag. 127–38.

Wehrsig, Christof, and Veronika Tacke (1992) "Funktionen und Folgen informatisierter Organisationen," in *ArBYTE. Modernisierung der Industriesoziologie,* Thomas Malsch and Ulrich Mill (eds.). Berlin: Edition Sigma.

Weinberger, Ota (1988) *Norm und Institution.* Vienna: Manzsche Verlag- und Universitätsbuchhandlung.

Weingart, Peter (2001) *Die Stunde der Wahrheit? Zum Verhältnis der Wissenschaft zur Politik, Wirtschaft und Medien in der Wissensgesellschaft.* Weilerswist: Velbrück Wissenschaft.

Wieacker, Franz (1970) "Zur praktischen Leistung der Rechtsdogmatik," in *Hermeneutik und Dialektik, Aufsätze II,* Rudiger Bubner, Konrad Cramer, and Reiner Wiehl (eds.). Tübingen: Mohr Siebeck Verlag. 311–36.

Winkler, Günter (1969) *Wertbetrachtung im Recht und ihre Grenzen.* Vienna: Springer Verlag.

Winter, Gerd, and Rüdiger Schäfer (1985) "Zur richterlichen Rezeption natur- und ingenieurwissenschaftlicher Voraussagen über komplexe technische Systeme am Beispiel von Kernkraftwerken," *Neue Zeitschrift für Verwaltungsrecht,* 4: 703–11.

Zweigert, Konrad (1969) "Rechtsvergleichung, System und Dogmatik" in *Festschrift für Eduard Bötticher zum 70. Geburtstag am 29. December 1969,* Karl August Bettermann and Albrecht Zeuner (eds.). Berlin: Duncker and Humblot. 443–9.

11

The Innovation Commons

Lawrence Lessig

The Internet revolution has ended just as surprisingly as it began. None expected the explosion of innovation and creativity that the network produced; few expected it to collapse as quickly and as profoundly as it has. The phenomenon has the feel of a shooting star, flaring unannounced across the night sky, then disappearing just as unexpectedly.

Yet neither the appearance nor disappearance of this revolution is difficult to understand. The difficulty is in accepting what a careful mapping teaches. For while the Internet was born in the United States, its success grew out of notions that seem far from modern American ideals of property and the market. Americans are captured by the idea, well expressed by Yale Law School professor Carol Rose, that "the world is best managed when divided among private owners," (1986: 711, 712) and when the market perfectly regulates those divided resources. But the Internet took off precisely because core resources were not "divided among private owners." Core resources of the Internet were left in a "commons." It was this commons that induced the extraordinary innovation that the Internet has seen. It is the enclosure of this commons that will bring about its demise.

This commons was built into the very architecture of the original network. Its design secured a right of decentralized innovation. It was this "innovation commons" that produced the diversity of creativity that the network has seen within the United States, but even more dramatically outside the United States. The potential for a free and open platform for commerce and innovation is of greatest benefit where the real space alternatives are neither free nor open.

Policy makers—especially those outside the United States—need to understand the importance of this architectural design to the innovation and creativity of the original net. The potential of the internet has just begun to be realized, especially outside the United States. Yet old ways of thinking are reasserting themselves within the United States to change this design. Changes to the Internet's original core will in turn threaten the net's potential everywhere—staunching the opportunity for innovation and creativity. Thus, just at the moment this transformation could have a real and meaningful effect, a counter revolution is succeeding in undermining the potential of this network

The motivation for this counter-revolution is as old as revolutions themselves. As Machiavelli described it long before the Internet: "Innovation makes enemies of all those who prospered under the old regime, and only lukewarm support is forthcoming from those who would prosper under the new." And so it is today with us: Those who prospered under the old regime are threatened by the Internet; this is the story of how they react. Those who would prosper under the new regime have not risen to defend it against the old; whether they will is still a question. The answer so far is that they will not.

The Neutral Zone

A "commons" is a resource to which everyone within a relevant community has equal access. It is a resource that is not, in an important sense; "controlled." Private or state owned property is a controlled resource; only as the owner specifies may that property be used. But a commons is not subject to this sort of control. Neutral or equal restrictions may apply to it (an entrance fee to a park), but not the restrictions of an owner. A commons, in this sense, leaves its resources "free."

Commons are features of any culture. They have been especially important to cultures outside the United States. Professor of political science Elinor Ostrom has done much to demonstrate their importance in a wide range of contexts across the world—from communal tenure systems in Switzerland and Japan to irrigation communities within the Philippines. But within the American intellectual culture, a commons is treated as an imperfect resource. It is the object of "tragedy," as biologist Garrett Hardin famously described; it provides an imperfect incentive to consume and deploy resources. Wherever a commons exists, the aim is to enclose it. Commons are unnecessary holdovers from time gone by, best removed if possible.

For most resources, for most of the time, this bias against the commons makes good sense. It is true that when some resources are left in common, individuals are driven to over consume the resource and therefore deplete it. But for some resources, this bias against the commons is blinding. Some resources are not subject to the "tragedy of the commons," as obviously, some resources cannot be "depleted." (No matter how much we use the resource of Einstein's Theory of Relativity, or copy Robert Frost's poem "New Hampshire," those resources will survive.) For these resources, the challenge is to induce provision, not to avoid depletion. But as the problems of provision are very different from the problems of depletion, to confuse the two is to misguide policy.

This misguidance is particularly acute when considering the Internet. At the core of the Internet is a design—chosen without a clear sense of its consequences—that was new among large scale computer and communication networks. Named the "end to end argument" by network architects Jerome Saltzer, David Clark, and David Reed, this design influences where "intelligence" in the network is placed. Traditional computer-communications systems located intelligence, and hence control, within the network itself. Networks were "smart"; they were designed by people who believed they knew everything the network would be used for.

But the Internet was born just at the time that a different philosophy was arising within computer science. This philosophy ranked humility above omniscience. It expected that the network designers would have no clear idea about all the ways the network could be used. It therefore counseled a design that built the minimum possible into the network itself, leaving the network free to develop as the ends (the applications) wanted.

The motivation for this new design was flexibility. The consequence was innovation. Because innovators needed no permission of the network owner before a different application or content got served across the network, innovators were much more free to develop new modes of connection. Technically, the network achieved this design simply by focusing upon the delivery of packets of data, without worrying about either the contents of the packets or their owners. Nor does it worry that all the packets make their way to the other side. The network is "best efforts"; anything more is provided by the applications at both ends. Like an efficient post office (imagine!), the system simply forwards the data along.

By refusing to optimize the network for any single application or service, the Internet kept the network open to new innovation. The World Wide Web is perhaps the best example. The Web was the creation of Tim Berners-Lee at the CERN laboratory in Geneva. Berners-Lee wanted to enable users on a network to get easy access to documents located elsewhere on the network. He therefore developed a set of protocols to enable hypertext links among documents located across the net. Because of end-to-end, these protocols could be layered on top of the initial protocols of the Internet. This meant that that the Internet could grow to embrace the Web. Had the network compromised its commitment to end-to-end—had its design been optimized to favor telephony, for example, as many in the 1980s wanted—then the Web would not have been possible.

This end-to-end design is the "core" of the Internet. If we can think of the network as built in layers, then the end-to-end design was effected through a set of protocols implemented at the middle layer—what we might call the logical, or code layer, of the Internet. Below the code layer is a physical layer (the computers and the wires that link the computers). And above the code layer is a content layer (the stuff that gets served across the network). Not all these layers were organized as commons. The computers at the physical layer are obviously private property, not "free" in the sense of a commons. And much of the content served across the network is copyrighted, and hence protected. It too is not "free" for all to take in the sense that a commons is.

But the code layer of the Internet was organized as a commons. By design, no one controls the resources for innovation that get served across this layer. Individuals control the physical layer, deciding whether a machine or network gets connected to the Internet. But once connected, at least under the Internet's original design, the innovation resources for the network remained free.

No other large-scale network left the code layer free in this way. For most of the history of telephone monopolies worldwide, permission to innovate on the telephone platform was vigorously controlled. In the United States, in 1956. AT&T successfully persuaded the U.S. Federal Communications Commission to block the use of a plastic cup on a telephone receiver, designed to block noise from the telephone microphone, on the theory that AT&T alone had the right to innovation on the telephone network (Huber, Kellogg, and Thorne 1999: 416). A similar control over equipment connected to the network existed across the world.

The Internet might have remained an obscure tool of government-backed researchers had telephone companies stayed this way. The Internet would never have taken off had ordinary individuals not been able to connect to the network by way of Internet service providers (ISPs) through already existing telephone lines. Yet this right to connect was not preordained. Instead, it is here that an important accident in regulatory history began to have a role. Just at the moment the Internet was emerging, the telephone company was being moved to a very different regulatory paradigm. Up until the early 1980s, the telephone monopoly was essentially free to control its wires as it wished. But beginning slowly in the late 1960s, and then much more vigorously through the 1980s, the government began to require that the telephone company behave neutrally: First, by requiring that the telephone companies permit customer premise equipment to be connected to the network, and then by requiring that the telephone companies permit others to have access to their wires.

This requirement was rare among telecommunications monopolies world-wide. In Europe and throughout the world, telecommunications monopolies were permitted to control the uses of their network. No requirement of access operated to enable competition. And hence no system of competition grew up around these other monopolies. But when the United States broke up AT&T in 1984, the resulting companies no longer had the freedom to discriminate against other uses of its lines. And hence when ISPs sought access to the local Bell lines to enable customers to connect to the Internet, the local Bells were required to grant access equally. This enabled a vigorous competition in Internet access, and this competition meant that the network could behave strategically against this new technology. In effect, through a competitive market, an end-to-end design was created at the physical layer of the telephone network, which meant that an end-to-end design could be layered on top of that.

This innovation commons was thus layered onto a physical infrastructure that, through regulation, had important commons-like features. Common-carrier regulation of the telephone system assured that system could not discriminate against an emerging competitor: the Internet. And the Internet itself was architected, through its end-to-end design, to assure that no particular application or use could discriminate against other, different innovations. Neutrality thus existed at the physical and code layer of the Internet.

And so too was there an important neutrality at the content layer of the Internet. As I have described, the content layer includes all the content streamed across the network—web pages, MP3s, e-mail, streaming video—as well as application programs that run on, or feed, the network. These programs are distinct from the protocols at the code layer—TCP/IP, including the protocols of the World Wide Web. These protocols are dedicated to the public domain.

But the code above these protocols is not in the public domain. It is instead of two sorts: proprietary and non-proprietary. The proprietary includes the familiar Microsoft operating systems and web servers, as well as programs from other software companies. The non-proprietary include open source and free software, especially the Linux (or GNU/Linux) operating system, the Apache server, as well as a host of other plumbing-oriented code that makes the net run.

This non-proprietary code creates a commons at the content layer. The commons here is not just the resource that a particular program might provide—for example, the functionality of an operating system or of a web server. The commons also includes the source code of this software that can be drawn upon and modified by others. Open source and free software ("open code" for short) must be distributed with the source code; the source code must be free for others to take and modify. This commons at the content layer means that others can take and build upon open source and free software. It also means that that open code can not become captured and tilted against any particular competitor. Open code can always be modified by subsequent adopters; it therefore is licensed to remain neutral among subsequent uses. There is no "owner" of an open code project.

In this way, and again, parallel to the end-to-end principle at the code layer, open code decentralizes innovation. It keeps a platform neutral. And this neutrality in turn inspires innovators to build for that platform because they need not fear the platform will turn against them. Open code thus builds a commons for innovation at the content layer. Like the commons at the code layer, open code preserves the opportunity for innovation, and protects innovation against the strategic behavior of competitors. Free resources thus induce innovation.

An Engine of Innovation

The original Internet, as it was extended to society generally, thus mixed controlled and free resources at each layer of the network. At the core code layer, the network was free. The end-to-end design assured that no network owner could exercise control over the network. At the physical layer—the computers and the wires that linked those computers—the resources were essentially controlled, but even here, important aspects were free. One had the right to connect a machine to the network or not, but the telephone companies did not have the right to discriminate against this particular use of their network. And finally, at the content layer, much of the resources served across the Internet were controlled.

But a crucial range of software building essential services on the Internet remained free. Whether through an open source or free software license, these resources could not be controlled.

This balance of control and freedom produced an unprecedented explosion in innovation. The power, and hence the right, to innovate was essentially decentralized. The Internet might have been an American invention, but creators from around the world could build upon this network platform. And significantly, some of the most important innovations for the Internet came from these "outsiders."

The most important technology for accessing and browsing the Internet— the World Wide Web—was not invented by companies specializing m network access. It was not AOL or Compuserve. The Web was developed by a researcher in a Swiss laboratory who first saw its potential and then fought to bring it to fruition. Likewise, it was not existing e-mail providers who came up with the idea of Web-based email. That was the creation of an immigrant to the United States from India, Sabeer Bhatia, and it gave birth to one of the fastest growing communities in history: Hotmail.

And it was not traditional network providers, or the telephone company, that invented the applications that enabled online chatting to take off. The original community-based chatting service (ICQ) was the invention of an Israeli, far from the trenches of network design. His service could explode (and then be purchased by AOL for $400 million) because the network was left open for just this type of innovation.

Finally, the revolution in bookselling initiated by Amazon.com through the use of technologies that "match preferences" of customers was invented far from the traditional organs of publishers. By gathering a broad range of data about purchases by customers, Amazon.com, drawing upon technology first developed at MIT, can predict what a customer is likely to want. Amazon can then offer the customer what he or she is likely to want. These recommendations thus drive sales, but without the high cost of advertising or promotion. Thus booksellers such as Amazon can out-compete traditional marketers of books, which may in part account for the rapid expansion of Amazon into Asia and Europe.

These innovations are at the level of Internet services. Far more profound have been innovations at the level of content. The Internet has not only inspired invention, it has inspired publication in a way that would never have been produced by the world of existing publishers. From collaborative data-bases collecting information about CDs, or movies, to online archives of lyrics and chord sequences—these are all instances of a kind of creativity that was enabled because the right to create was not controlled.

Again, the examples have not been limited to the United States. OpenDemocracy.org is a London based web-centered forum for debate and exchange about democracy across the world. It promises to become the first international forum for international exchange about ideas relating to democracy and governance. Such a forum could be born only because no coordination among international actors was needed. And it can thrive because as at a low cost it can engender debate among different views.

This history should be a lesson. Every significant innovation on the Internet has emerged from outside of traditional providers. The new grows away from the old. This teaches the value in leaving the platform for innovation open. But unfortunately, that platform is now under siege. Every technological disruption creates winners and losers. The losers have an interest in avoiding that disruption if they can. This was the lesson Machiavelli taught; it is the experience with every important technological change over time. It is also the experience we are now seeing with the Internet. The innovation commons of the Internet threatens important and powerful pre-Internet interests. During the past five years, those interests have mobilized to launch a counterrevolution that is now having a global impact.

This movement is fueled by pressure at both the physical and content layers of the network. These changes, in turn, put pressure on the freedom of the code layer. These changes will have an effect on the opportunity for growth and innovation that the Internet presents. And policymakers keen to protect that growth should be skeptical of changes that will threaten it. Broad-based innovation may threaten the profits of some existing interests, but the social gains from this unpredictable growth will far outstrip the private losses—especially in nations just beginning to connect.

Fencing Off the Commons

The Internet took off on the telephone lines Narrowband service across acoustic moderns enabled millions of computers to connect through thousands of ISPs. Local service providers had to provide ISPs access to local wires; they were not permitted to discriminate against Internet service. Thus the physical platform on which the Internet was born was regulated to remain neutral. This regulation had an important effect.

A nascent industry could be born on the telephone wires regardless of the desires of the telephone company.

But, as the Internet moves from narrowband to broadband, this feature of the initial regulatory environment is changing. The dominant broadband technology in the United States is currently cable. Cable lives under a different regulatory regime. Cable providers in general have no obligation to grant access to their facilities; there is no obligation not to discriminate in the service they provide. Thus cable has asserted the right to discriminate in the Internet service it provides.

Consequently, cable has begun to push for a different set of principles at the code layer of the network. Cable companies have deployed technologies to enable them to engage in a form of discrimination in the service that they provide. Thus, Cisco, for example, has developed "policy based routers" that will enable a cable company to choose which content flows quickly, which flows slowly. With these, and other technologies, the cable companies will be in a position to exercise power over the content and applications that operate on their network.

This control has already begun within the United States. ISPs running cable services have exercised their power to ban certain kinds of applications (in particular, those that enable peer-to-peer service.) They have blocked particular content (advertising from competitors, for example) when that content was not consistent with their business model. The model for these providers is the model of cable television generally—controlling access and content to the cable providers' end.

The European experience has been somewhat different though these ideals of increasing control are slowing working their way across the Atlantic. Broadband penetration in Europe is quite limited. There was less than 0.25 broadband connections per 100 inhabitants in Germany in 2000; the United States enjoyed ten times that. (South Korea, meanwhile, had 40 times the broadband connectivity of Germany in 2000.) But the initial push of the European Union has been to increase competition in telecommunications, as a means to spur Internet growth.

The initial moves have mirrored die policy of neutrality that was pressed in the United States at the start. But increasingly, a counter-movement in Europe is echoing the same ideas of control from the United States. Once again, the notion that network owners exercise control at the code layer (and hence control over content) is gaining.

To the extent cable becomes dominant mode of access to the Internet, this will change the environment of innovation existing on the original network. Rather than a network that vested intelligence in the ends, this network will vest an increasing degree of intelligence within the network itself. And to the extent it does this, the network will increase the opportunity for strategic behavior in favor of some technologies and against others. An essential feature of neutrality at the code layer will have been compromised. This compromise will reduce the opportunity for innovation for those outside the United States especially.

Far more dramatic, however, has been die pressure from die content layer on the code layer. This pressure has come in two forms. First, and most directly related to the content I described above, there has been an explosion of patent regulation in the context of software. Second, there has been the increasing control that copyright holders have exercised over new technologies for distribution.

The changes in patent regulation are the more difficult to explain, though the consequence is not hard to track. Beginning two decades ago, the United States patent office began granting patents for software-like inventions. In the late 1990s, the court overseeing these patents finally approved the practice, and approved their extension to "business methods." (The European Union, meanwhile, initially adopted a much more skeptical attitude towards software patents. But the pressure from the United States will eventually bring the European Union's policy into alignment with the American policy.)

In principle, these patents are designed to spur innovation. But with sequential and complementary innovation, there is almost no evidence that such patents will do any good, and increasing evidence that they will do actual harm. Like any regulation, patents tax the innovative process generally. Like any tax, some (large firms rather than small. U.S. firms rather than foreign) are better able to bear that tax than others. Open code projects, in particular, are threatened by this explosion in patents, as they are least able to negotiate appropriate patent licenses

The most dramatic restrictions on innovation, however, have come at the hands of copyright holders. Copyright is designed to secure to artists control over their "writings" for a limited time. The aim is to secure to copyright holders a sufficient interest to produce new work. But copyright laws were crafted in an era long before the Internet. And their effect in the context of the Internet has been to transfer control over innovation in distribution from innovators generally to a relatively concentrated few.

The clearest example of this effect is online music. Before the Internet, the production and distribution of music had become extraordinarily concentrated. In 2000, for example, five companies controlled 84 percent of music distribution in the world. The reasons for this concentration are many— including the high costs of promotion— but the effect of concentration on artist development is profound, very few artists make any money from their work, and the few that do are able to do so because of mass marketing from the record labels. The Internet had the potential to change this reality. Both because the costs of distribution were so low, and because the network had the potential to lower the costs of promotion significantly as well, the cost of music could fall, and the revenues to artists could rise.

Thus five years ago, this market took off. A large number of online music providers began competing for new, innovative ways to distribute music. Some distributed MP3s for money (eMusic.com). Some built technology for giving owners of music easier access to their music (MP3.com). And some made it much easier for ordinary users to "share" their music with other users (Napster).

But as quickly as these companies took off, lawyers representing old media succeeded in shutting them down. These lawyers argued that copyright law gave the holders (some say hoarders) of these copyrights the exclusive right to control how they get used. American courts agreed.

Now to keep this dispute in context, we should think about the last example of a technological change that facilitated a much different model for distributing content: Cable TV which has been accurately hailed as the first great Napster. Owners of cable television systems essentially set up antennae and "stole" over-the-air broadcasts, and then sold that "stolen property" to their customers. But when the courts were asked to stop this "theft," they refused. Twice the Supreme Court held that this use of someone else's copyrighted material was not inconsistent with the copyright law.

When the U.S. Congress finally got around to changing the law, it struck an importantly illustrative balance. Congress granted copyright owners the right to compensation from the use of their material on cable broadcasts, but cable companies were given the right to broadcast the copyrighted material. The reason for this balance is not hard to see. Copyright owners certainly are entitled to compensation for their work. But the right to compensation should not translate into the power to control innovation. So rather than giving copyright holders the right to veto a particular new use of their work (in this case, because it would compete with over-the-air broadcasting). Congress assured copyright owners would get paid without having the power to control. Compensation without control.

The same deal could have been struck by Congress in the context of online music. But this time, the courts were not hesitant to extend the copyright holders control. So the concentrated holders of these copyrights were able to stop the deployment of competing distributors. And Congress was not motivated to respond by granting an equivalent compulsory right. The aim of the recording company's strategy was plain enough: To shut down these new, and competing models of distribution and then replace them with a model for distributing music online more consistent with the traditional mode.

This trend has been supported by the actions of Congress. In 1998, Congress passed the Digital Millennium Copyright Act (DMCA), which (in)famously banned technologies designed to circumvent copyright protection technologies, and created strong incentives for ISPs to remove any material from their sites claimed to be a violation of copyright.

On the surface both changes seem sensible enough. Copyright protection technologies are analogous to locks. What right does anyone have to pick a lock? And ISPs are in the best position to assure that copyright violations do not occur on their web sites. Why not create incentives for them to remove infringing copyrighted material?

But intuitions here mislead. A copyright protection technology is just code that controls access to copyrighted material. But that code can restrict access more strongly (and certainly less subtly) than copyright law does. Thus, often the desire to crack the protection systems is nothing more than a desire to exercise what is sometimes called a fair-use right over the copyrighted material. Yet the DMCA bans that technology, regardless of its ultimate effect.

More troublingly, it is effectively banning that technology on a worldwide basis. When the Russian programmer Dmitry Skylarov, for example, wrote code to crack Adobe's eBook technology, it was to enable users to move eBooks from one machine to another, or blind consumers to use technology to "read" the books they purchased out loud. The code Skylarov wrote was legal where it was written. When it was sold by his company in the United States, it became illegal. When he came to the United States to talk about that code, the FBI arrested him. Today he faces a sentence of twenty-five years for writing code that could be used for fair use purposes but also to violate copyright laws.

The same trouble has arisen with the provision that gives ISPs the incentive to take down infringing copyrighted material. When an ISP is notified that material on its site violates copyright it can avoid liability if it removes that material. As it does not have any incentive to expose itself to liability, the ordinary result of such notification is for the ISP to remove the material. Increasingly, companies trying to protect themselves from criticism have used this provision to silence critics. In August, for example, a British pharmaceutical company invoked the DMCA to get an ISP to shut down an Animal Rights site that was criticizing the British company's. Said the ISP, "it's very clear [the British company] just wants to shut them up," but ISPs have no incentive to resist the claims (Mieszkowski, "No free speech for animal rights web sites," *salon*, 31 August 2001).

Some believe this form of American regulation will only directly affect Americans. But the expanding jurisdiction that American courts claim, tied with the push by the World Intellectual Property Organization to enact similar legislation elsewhere, means that the effect of this sort of control will be felt world wide. There is no "local" when it comes to corruption of the Internet's basic principles.

In all these cases, there is a common pattern. In the push to give copyright owners perfect control over their content we also give copyright holders the ability to protect themselves against innovations that might threaten existing business models. The law becomes a tool to assure that new innovations do not displace old—when instead, the aim of the copyright and patent law should be, as the United States Constitution requires, to "promote the progress of science and the useful arts."

The effect of these changes will be felt globally, and perhaps most profoundly, outside the United States. To the extent the changes weaken the open source and free software movements, countries with the most to gain from a free and open platform lose. Those include the nations in the developing world, and those nations who would not want to cede this crucial architecture of control to a single private corporation. And to the extent that content becomes more controlled, then nations that could benefit from vigorous competition in the delivery and production of content will lose. An explosion of innovation to deliver MP3s would directly translate into innovation to deliver telephone calls and video content. Lowering the cost of this would dramatically benefit nations that still suffer from weak technical infrastructures.

Policy-makers around the world must recognize that the interests most strongly protected by this counter revolution are not their own, that resisting these changes to the environment of innovation offered by the Internet would better protect the opportunity the Internet affords them. And most importantly, they should be skeptical of press for legal power to enable those most threatened by the innovation the Internet promises to resist.

The Internet promised the world—and most significantly, the weakest in the world—the fastest and most dramatic change to existing barriers to growth. That promise depends upon the network remaining open to innovation. That openness depends upon policy that better understands the Internet's past.

References

Huber, Peter W., Michael K. Kellogg, and John Thorne (1999) Federal Telecommunications law. 2nd edition. Gaithersburg, Md.: Aspen Law and Business. 416.

Rose, Carol (1986) "The comedy of the commons: Custom, commerce, and inherently public property," *University of Chicago Law Review* 53: 711–2.

12

Quasi-Science and the State: 'Governing Science' in Comparative Perspective

Stephen Turner

Open discussion is a political ideal that today stands atop the hierarchy of the common political values of the West, and of the world at large. Its nineteenth-century origins are to be found in James Mill, who took it from the physiocrats of the century before. As John Stuart Mill put it in his autobiography, "so complete was my father's reliance on the influence of reason over the minds of mankind, whenever it is allowed to reach them, that he felt as if all would be gained if the whole population were taught to read, if all sorts of opinions were allowed to be addressed to them by word and in writing, and if by means of the suffrage they could nominate a legislature to give effect to the opinions they adopted" (cited in Dicey [1914] 1962: 161–2). As John Austin said, the elder Mill "held, with the French *Economistes,* that the real security of good government is in a *peuple éclairé*" (quoted p. 163). The language reappears in Habermas, as does the sentiment (1973: 282).

Is a "*peuple éclairé*" a meaningful ideal today in the face of the complexity of expert knowledge? In *Risk Society*, Ulrich Beck (1992) applies it to the problem of experts, but modifies it in a significant way. Beck gave voice to a demand for "forums of consensus-building cooperation among industry, politics, science, and the populace" (p. 29). This amounts to a rejection of the conventional electoral legislative representation mentioned by Mill. Subjecting experts to the discipline of consensus-building is a form of governance, and it is perhaps trivially true that even science is always governed through some sort of consensual, negotiated cooperation between these elements. But it is also true that the notions of cooperation and consensus could be specified in a number of divergent ways. Beck has a maximalist view of the nature of the appropriate consensus; he argues that this demand can only be met if certain conditions are satisfied:

> First, people must say farewell to the notion that administrations and experts always know exactly, or at least better, what is right and good for everyone: demonopolization of expertise.
>
> Second, the circle of groups to be allowed to participate can no longer be closed according to considerations internal to specialists, but must instead be opened up according to *social* standards of relevance: informalization of jurisdiction.
>
> Third, all participants must be aware that the decisions have not already been made and now need only to be "sold" or implemented externally: opening the structure of decision-making.
>
> Fourth, negotiating between experts and decision-makers behind closed doors must be transferred to and transformed into a public dialogue between the broadest variety of agents, with the result of additional uncontrollability: creation of a partial publicity.
>
> Fifth, norms for this process—modes of discussion, protocols, debates, evaluations of interviews, forms of voting and approving—must be agreed on and sanctioned: self-legislation and self-obligation. (p. 29–30)

This list points to a model of governance: stakeholders meeting in negotiation forums. Beck says that there is no assurance that these measures would work: "negotiation forums are not consensus production techniques with a guarantee of success." However, they can "urge prevention and precaution and work towards a symmetry of unavoidable sacrifices" (p. 30).

Can any institutional framework of the sort implied by Beck's "demand" solve the underlying problem with experts: that expert knowledge is simply inaccessible to non-experts, who are forced, like it or not, to accept the authority of experts? Is the problem one of institutional form, or is the problem the nature of the knowledge in question? The problem of "the governance of science" in the narrow sense, the problem of the intelligent exercise of popular control over science, is a simulacrum of this larger problem. If those who are empowered to make decisions about science cannot understand the science they are making decisions about, or perhaps even meaningfully communicate about it, is the intelligent exercise of popular control over science possible?

As a matter of political theory, the problem is fundamental to democracy. It was not a left-wing ruler, but Harry Truman who affirmed the following, in his vetoing of the original bill to establish the National Science Foundation:

> This bill contains provisions which represent such a marked departure from sound principles for the administration of public affairs that I cannot give it my approval. It would, in effect, vest the determination of vital national policies, the expenditure of large public funds, and the administration of important governmental functions in a group of individuals who would be essentially private citizens. The proposed National Science Foundation would be divorced from control by the people to an extent that implies a distinct lack of faith in democratic processes. (Penick 1965: 135)

There is a problem of "governance" of science because science is not self- funded. Ordinarily it must rely on "governmental" mechanisms' real states, or state based authorities, and the money of the populace and the industries from which taxes are extracted. And this enables us to specify the problem more clearly. Democracies require accountability, and as a practical matter, the problem of accountability requires a bureaucratic solution or authority that can be held responsible.

Bureaucratic solutions, like constitutions, do not travel well, and cannot be simply transplanted into a new setting. Thus the problem of national bureaucratic and governmental traditions is closely related to the problem of the governance of science in the sense of accountability: Traditions serve both to define the form of politics about expertise, and to determine what forms of "governance of science" are practically feasible in a given national political tradition. So what I will have to say here will have a great deal to do with a comparative problem. And what I will argue is that different national administrative traditions have led to different forms of the governance of science.

For convenience we can restrict our discussion to liberal democracies. But it is important to see that the problem is one that arises in a particular form as a result of the transition from non-democratic to democratic orders. Beck's maximalist solution to the problem of expert knowledge is to bring it to the level of bargaining between stakeholders. Traditionally, science has occupied a different category, and as Beck sees it a privileged or monopolistic category. Yet it would be a mistake to think that the creation and acceptance of such a status is inherently "anti-democratic." It would of course be possible to have a democratic consensus that agreed to assign the task of supporting and controlling science to the category of national security, and locate control in the military, which is traditionally exempt, or largely so, from political control, as it was formally in the Wilhelmine period in Germany. The military, indeed, represents a traditional form of expertise, together with a form of accountability. Military control was a model that the United States (and also of course Germany and the United Kingdom) operated with in connection with atomic research during the war and, in the case of the United States, well into the Cold War. But this model was thought to be a temporary expedient, inconsistent with the requirements of science itself, namely its requirements of open discussion, and thus inapplicable to other areas of science and, in the long run, to nuclear research itself. But it signals that there are radically different possible solutions to the problem of accountability, and to the problem of what administrative structures are appropriate for "governing" science in the external sense. But while the form of delegation is equally appropriate to monarchs or democratic regimes, a specific problem arises with the delegation of powers in democracies, and Beck's discussion allows us to pinpoint it.

Delegation

The standard solutions to the problem of governance in the external sense recognize the inability of public discussion of the sort "demanded" by Beck to deal effectively with science. This places the problem of science in a larger class of problems for which it is recognized that public discussion is inadequate, and for which it is reasonable as a matter of efficiency, justice, expedience, or some other consideration to delegate the task to others. Delegation often occurs when there are strong public passions of one sort or another that legislatures do not wish to take responsibility for or believe simply to be misguided. Monetary policy, for example, in most countries, is delegated to specialists who are positionally neutral, such as governors of reserve banks. These figures are also selected because of their expertise. But the tasks that they perform are not "scientific" economic tasks, though they may well act on the basis of economic evidence and accepted economic principles. Their task is essentially a political one—and one granted because the task cannot be trusted to ordinary legislative representative politics (similarly for judges). And, in a more complex way, bureaucracies are forms of delegation—forms that have traditionally fit very problematically with democracy, in part because of their historical origins as delegations of *royal* powers, and as an executive extension of the royal staff.

Presumably Beck—who is of course not a libertarian—has some notion of delegation, to the executive powers of the state and its bureaucracies. So the issue is not whether to delegate, but what to delegate and how to delegate. If one is going to argue that the present modes of delegation are inappropriate political models for dealing with technical decisions, one must deal with the problems that these techniques are designed to solve: the necessity of at least minimal technical competence and specialized knowledge on the part of the decision makers, something ordinarily not shared with representatives to parliament or with ordinary generalist civil servants. One must also question whether the quality of the decisions would be worse if they were subject to ordinary politics. It is, so to speak, a constitutional question as to whom a given decision ought to be entrusted—to representatives directly, to "the people," to a "cooperative" arrangement, or to a specialized body.

In legal contexts, we are perfectly comfortable with the notion that considerations of legal guilt and innocence ought not to be simply the province of representatives of the people. The history of the tribunes in the Roman legal system and the failings of lynch law are ordinarily understood to be good grounds for creating an insulated specialist judiciary, and in addition, in the Anglo-Saxon tradition, the practice of delegating the tasks of fact-finding as distinct from matters of law to juries selected from the people. The legal analogy is intriguing, because it poses the problem quite directly. Who is to decide what is a question of fact and what is a question of law? One might imagine a system in which juries decided what their own competence was, and could ask judges at their discretion for advice about matters of law. But the system is one in which judges decide what is a matter of law and what is not.

Forms of Delegation

The relevance of these examples is this: there are many ways of delegating limited powers of the state, or of dividing up discussions, decision-making powers, advisory roles, and so forth. Other forms of expertise, closer to that of science proper, have a long history of delegation as well, notably in the medical sphere, where state authority was used to deal with epidemics. In addition, there are procedures for licensing and certifying professionals. The ubiquity of these procedures raises an important point Even Beck presumably would not advocate the abandonment of professional licensure. And there seems to be no argument to the effect that delegation as such m relation to technical questions is a bad thing, nor does it seem that there is a conflict in principle with either the principles of liberal "government by discussion" or democracy with a (democratic liberal) decision to delegate powers to an expert or mixed body. But these all represent forms of the governance of science. There are issues of conflict of interest and particular problems of trust that arise in connection with particular forms of delegation. Nevertheless, there does seem to be a tendency, evident in Beck, to idealize some sort of participation or participatory democracy as the appropriate solution to problems with experts. The difficulty with this solution, as I have hinted, is that there seems to be no coherent type of discourse that could be engaged in by members of the public as such communicating and arguing directly with "experts." Indeed, most of the commentators on these issues are loath to suggest that conventional forms of democracy be granted special powers over experts, or to seize responsibility for technical decisions or technical controversies in general. Indeed, one of the *reductio ad absurdum* arguments against state control of science in the late 1930s and early 1940s in Britain pointed to the possibility that the choice of reagents in a chemistry lab would be subject to popular vote. What is ordinarily called for is something else: some novel kind of forum in which experts could be in some sense held accountable or forced to legitimate themselves. Beck's demands of course go beyond this, but in the same direction—towards some sort of popular control of science.

Accountability and legitimacy are terms of political theory that refer to different mechanisms of popular control. Accountability ordinarily means that there is a greater degree of supervision, usually based on numerical targets and procedural guidelines. Legitimacy is the idea that the actions of political leaders or institutions are accepted as valid. One of the oddities of the discussion of these issues in Beck and Brian Wynne (1996) is their apparent assumption that these processes of legitimation and response to the public are not in fact ongoing, and that there is some special crisis of legitimacy that is in particular associated with "experts," as distinct from problems of accountability. Discussions of "legitimation crisis" of course have long been a staple of critical theory and the idea that the political regimes of the West are constantly skating at the edge of de-legitimation is an important part of the imagery employed by these writers. But legitimacy in this sense is extraordinarily hard to assess.

The issue of legitimacy, or some analogue to it like "trust in the claims of experts," arises routinely in American liability cases; in that context commentators routinely argue that "junk science" is given too much credibility and that the motives and conflicts of interest of experts and claimants are given too much weight. Indeed Sheila Jasanoff herself seems to endorse this claim in her commentary on the courts (1995). The distrust takes a conventional form in these cases. Large corporations are considered to be able to pay (and are criticized if they don't pay), and not only capable of lying, but eager and willing to lie, and also able to hire "experts" who will lie on their behalf. This certainly does not speak to any sort of problem of experts being taken too seriously. And indeed scientists commenting on the role of junk science in the courtroom routinely complain that jurors and judges are not adequately cognizant of the importance of the reputation for institutions such as universities, and the unlikelihood that expert witness research done by them would be tainted by the fact that it is paid for by large corporations in the midst of bitter disputes.

The issues here are complex, and so many interrelated issues arise in each given court case that it is difficult to say what role expert witnesses play. Nevertheless, jurors are clearly capable of discounting expert claims on the grounds of what, for them, amount to conflicts of interest arising from the fact that they make their living as expert witnesses. This scepticism is entirely justified by the incredibly seamy practices of attorneys using expert witnesses. Expert witnessing is a business with an active market and active marketing. Any personal injury lawyer in the United States is deluged with advertisements from "expert witnesses" seeking work. These witnesses are clearly concerned about the marketability of their particular strengths, such as apparent candor and sincerity, and so are the lawyers who hire them. In an adversary system this presumably is the way it should be. But, plaintiff lawyers are also successfully tapping into an enormous amount of suspicion of the motives and credibility of large corporations and of the scientists they employ, however apparently independent they might be. This suggests that the whole issue of the credibility and power of experts, as much as the literature wishes to separate it, cannot be separated from the particular institutions in which expert witnessing or expertise plays a role and the general experiences and attitudes of particular audiences with these institutions. But negative experiences with institutions of delegated powers are common, and not restricted to science.

In the United States, for example, many people have negative experiences with the federal government and federal agencies, which they perceive as arrogant, irrational, punitive, and self-serving. It would be astonishing if scientists employed by these same agencies were thought to be credible, honest, and truth telling, and indeed they are not. The different configurations of expert distrust in European countries almost certainly reflect the distinctive experiences of citizens of those countries with their own civil service bureaucrats, politicians, regulators: in the case of Britain, a special civil service dominated by generalists, and in the case of Germany, a bureaucracy dominated by secrecy and hidden discretion. The political decision to grant discretionary powers to these kinds of agencies is subject to the consequence that each bureaucratic form creates its own suspicions. In the United States, the suspicion that government agencies are attempting to make work for themselves often undermines their credibility, as does the fact that people experience the arbitrariness of government bureaucrats enforcing unintelligible regulations in an irrational way.

To be sure, citizens ordinarily do not grasp the details of organizational structure that motivates particular forms of bureaucratic behavior. Nevertheless they get a general impression about the variety and motivation of bureaucrats and act accordingly. The same holds in continental and British contexts, though the distribution of authority is significantly different and the forms that distrust take are accordingly also different. If we looked at the responses of the public as rational responses to the kind of information that they acquire in ordinary context about these institutions and about experts, we would see a learning process by which citizens change their views of the utterances of bureaucrats, especially on such subjects as risk when these utterances are confounded by events.

If we consider the problem of delegation in its earlier contexts—and this is oddly preserved in the enlightenment conception of the public— the delegation of the powers of the king are delegations of his own powers, powers that he could exercise if he was so inclined. Military powers are the paradigm of this—kings were warriors, like their vassals. By the time that Bismarck could tell Moltke that "the king has a great military mind—you," this was a dated illusion, but nevertheless not an incredible one. The idea that a king was a supreme legal authority died slowly, carried on in legal theory as the fiction that the law represented the will of the sovereign. But in the case of science, the illusion could not survive the first moments of the scientific revolution—that the king was the greatest of scientists is an absurdity not even the most convinced of monarchists could have voiced (it remained to the admirers of Stalin to do so). But for the democratic liberal successors to royal power, the problem was more acute. The claim that the public was "the great scientist" or the ultimate arbiter of scientific truth was too difficult to state (although Hobbes and the Enlightenment each did so with peculiar qualifications; Peters 1956: 57). And even in the age of kings there was a specific form of delegation that recognized the special character of scientific knowledge, and created appropriate forms for it—Royal Academics and prizes, such as the Nobel, which is awarded by a "royal" rather than a "parliamentary" body. The persistence of these monarchical forms is revealing—tension between the democratic and the scientific has not disappeared.

The Fragility of Expertise

The radical change in public opinion following the Three Mile Island nuclear disaster reflected new information. To understand this process of learning however it is quite useless to focus on the discrepancies between expert and public opinions on various subjects and hold them out as evidence of the irrationality of public opinion. Public opinion can of course be misled about risks and indeed *is* misled when the media report heavily on minor risks. Nevertheless the number of media reports and their apparent significance are data for the public, and it is this data that the public operates on, in conjunction with a great deal of other information, such as the information they derive from experience in dealing with government agencies about the motivations and competence of these agencies.

That experts are constantly having their acceptance as experts, their legitimacy, tested and questioned is a central feature of the notion of expertise and the phenomenon of delegated powers. The history of particular forms of expertise, such as the expertise of agricultural science, shows quite clearly that establishing credibility is difficult. Consumers are perfectly capable of rejecting the excuses given by experts. Selling expert knowledge is an extremely difficult matter, and conspicuous public successes are an important part of the process of gaining acceptance. A few snippets of this history will suffice to make this point. When Leibig first introduced commercial chemical fertilizers with the backing of "science," the fertilizers had various unexpected and unintended side effects that harmed farmers—this was among the first of the science-made environmental disasters. When the farmers were harmed they became extremely suspicious of the claims of fertilizer companies. Out of this came the demand for improved expertise, particularly with respect to soil chemistry, and state geologists eagerly stepped into the breach to provide improved information and means of testing soil chemistry.

Some farmers simply resisted change and ignored the blandishments of experts. These farmers became the focus of a lengthy public effort to improve agricultural practices and thus improve the lives of those farmers. The history of this enormous undertaking, chronicled in *The Reluctant Farmer* (Scott 1970), led to the creation of the Cooperative Extension Service in the United States, which sponsored such things as high school tomato-growing contests, cooking contests and the like in a more or less successful attempt to both involve poor farmers in the broader community and to get them to think more effectively about improvements in agricultural technique and the use of the results of agricultural science.

The interesting thing about this process is the incredible difficulty and fragility of expertise. The whole notion of an experimental farm subsidized by the state was for many farmers itself risible, and the fact that experimental farms required subsidies seemed to them to prove that the farms were bad farms. It was only through a wide variety of public efforts, such as the exhibition of prize animals, that schools of agriculture in the nineteenth century were able to impress upon farmers that agricultural scientists indeed did know something more that would be useful to them. But in most cases it was only the role of scientists in dealing with animal epidemics that really established the credibility of state university agricultural science. The results in these cases were visible, immediate and economically significant. To save a herd of cattle or a flock of sheep by relying on the advice or medicine of the agricultural scientist of the state university was to be seduced by experts. But it should not be thought that the experts had an easy time of time of this seduction or that farmers or anyone else were initially impressed by the claims of agricultural scientists.

What needs to be remembered is that the road to expertise and professionalization goes both ways. Groups can establish themselves as possessing expertise on the basis of either exemplary achievements or apparent successes in dealing with problems regarded as intractable, as was the case with the expertise of the various social movements of the late nineteenth century concerned with reform of prisons, juvenile justice, the treatment of fallen women, and the like. But the public can become dissatisfied with the results of this expertise and can decide to rely on other forms of action. The increase in juvenile crime, for example, has served to undermine the credibility of expert juvenile crime specialists who administer a juvenile justice system that was reformed in the early twentieth century on the basis of expert claims by various judges and reformers. The task of the juvenile detention halls and the like has in some cases been handed over to groups operating military-style boot camps and in other cases even to groups attempting to teach such things as Aristotelian ethics to juvenile prisoners. The message in these cases is simple. There are alternatives to professionalization and the "public" is perfectly capable of exerting pressure on political institutions to employ these alternatives.

Focusing on the example of science and its expertise is in fact somewhat narrow and misleading about the universe of expertise claims as a whole. Science and medical science earned an enormous amount of trust on the basis of such exemplary achievements as the discovery of penicillin, the polio vaccine, and the atomic bomb. But there is no reason in principle or in practice that this trust may not be dissipated, or that attempts to extend or use the authoritative influence of science in support of the tobacco lobby or extreme environmental regulation will not lead to the rejection of these forms of expertise.

What is the Governance of Science?

These cases yield complex lessons, but one lesson stands out. Powerful bureaucracies are, almost necessarily, conformist with respect to the expert opinions that they develop. Bureaucracies are not debating societies. They operate efficiently only when commands are followed according to rules and in accordance with limited discretionary power and in which career advancement depends on conformity to the culture of the bureaucracy. Bureaucracies are not prone to entertaining a diversity of viewpoints. Indeed, to delegate decision-making power or to delegate implementation of a policy to a large and powerful bureaucracy is almost necessarily to accept the particular expert opinions to which the bureaucracy conforms. This is not inconsistent with the doing of science, and indeed there is a great deal of bureaucratically organized science. But bureaucratic organization within science is a different matter from the bureaucratic organization *of* science itself.

Science, in contrast to bureaucracy, *is* a debating society. But even debating societies need to be governed, and to make decisions. If we consider the problem of decision making in the domain of quasi science as a separate and distinct problem, we can begin to ask some questions about how governance ought to be organized, especially to deal with questions of what sort of models of the organization of discourse and decision-making are available. Even the most cursory inquiry into this topic will show that science and quasi-science are already largely subject to various forms of "governance," including forms of self-governance. In science, as with other debating societies, governance characteristically begins with the problem of membership and the problem of regulating participation in discussion. It is obviously necessary to understand these governance structures, and the many more complex governance structures that depend on them, before asking questions about the redesign of the governance of science. Beck's model of discussion ignores these questions for a simple reason. His aim is to obliterate inequalities in status in discussions of this sort, which is precisely what most of the means of governance are designed to either establish or recognize.

The fault here is simple—this takes an abstract principle and treats it as an organizational directive. Like Hobbes' insistence that the king is the final arbiter of truth—something that makes sense in a specifically religious context as a part of the "more fundamental" task of the king to secure peace—it makes little sense in the practical light of existing complex institutions of delegation that have grown and established themselves through testing. Some of these means are so commonplace and ubiquitous as to escape notice.

To take a particularly obvious example, consider the licensing of new professionals in medicine, which is the largest "scientific" activity. Physicians are characteristically subject to at least three major forms of governance. The obtaining of medical qualifications is itself done through procedures at least partially controlled by qualified physicians, for example in the granting of medical degrees, qualifications that may in turn be subject to state control, as when the state regulates the granting of degrees, and in a federal system, where the national government either controls the accreditation of universities, of degrees, or accrediting agencies, as is the case in the United States. The state also may directly act to license physicians. And finally, the state traditionally exercises, among its police powers, powers over public health, particularly in the face of epidemic disease, and employs physicians as experts, as when making decisions to guarantee in which the physicians themselves exercise or have a privileged advisory role recognized by the state in the exercise of the police powers of the state.

Similar minimal combinations of these three governance roles or variations on them occur in many other expert and particularly scientific professions. If one thinks of a few of the relevant examples, such as the role of clinical psychologists in advising the state with respect to treatment of the mentally ill and the similar role of the social worker with respect to the care of children, one can see how these mechanisms have developed over time and indeed substantially expanded. One might even argue that the rise of these professions in particular and the learned professions generally is closely bound up with the phenomena of their governance, especially in those cases in which the professional exercises delegated powers of the state. Yet these professions are also partly, and in some cases largely, marketized. Patients may choose physicians, or, as is more common, institutions choose to employ physicians, social workers, and psychologists, who compete for positions. This competition is itself a form of control. So a large part of the "cooperation" between the populace, industry, and so forth that Beck is concerned with occurs through the mechanism of the market, but a market that is itself regulated in various ways, such as licensure.

With respect to science itself, the primary issue has traditionally been one form of the delegation of state power, the spending of the state's money to pursue the goals of science. This delegation of state power seems peculiarly benign compared with the exercise of police authority discussed above, but it nevertheless may have very substantial implications and consequences. The "governance," however, is characteristically indirect, and aspects of it are marketized. By "indirect" I mean what in political theory is known as the liberal preference for indirect forms of rule: forms of rule other than command and control. Beck's model implicitly relies on a scheme of command in the last instance—it involves *forced* cooperation, forced by the state acting as the agent of the populace, to control and replace the uncontrolled "cooperation" of the market. In science, it is a market that is already substantially governed by indirect means, and in which there is already competition for research funds.

The liberal preference for indirect means is a preference for non-coercive means, and the Left has traditionally criticized this preference as insufficient to bring about the results the state ought to be bringing about. But there is a deep problem with this critique: it rests on a questionable notion of the capacity of the state to bring about good results. The questions that are ordinarily raised are not merely a matter of liberal and Left, however. They reflect national experiences with particular means. The preference for marketizing, cooperative, or more direct forms, such as "planning," reflect experiences with each of these modes. The degree of federalization, the devolvement of powers to lower levels of government that can enter the market as buyers of science, or academic talent, varies as well. We will shortly return to these issues by considering the example of cholera in the nineteenth century. But there is a generic issue that needs to be addressed, that relates especially to science, over the character of decision-making and the existence of conflicts of interest, that arise from the fact that they *are* governed, however indirectly, or to put it differently, incompletely marketized.

The problem we face with decision-making in this domain is that it is quasi-scientific, in a particular sense: competent decision-making requires scientific knowledge, but scientific knowledge is not sufficient to make the decisions. The problem of conflict of interest arises because scientists are used to making decisions, or to giving advice to those who make decisions. Scientists are, to some extent at least, independent agents with career strategies independent of the state who may greatly benefit by acting for their own interest and against state interest, for example, by spending money on projects that advance their careers without advancing state purposes on which the state planned to spend the money. So decision-making must either be organized in such a way that conflicts of interest are avoided, or the risks of conflicts of interest ignored. The problem can be put most starkly by considering the most extreme model of the centralization of state scientific authority—one individual having the power to grant all of the positions that would enable a scientist to pursue research. Could this individual be trusted? It is questionable that any individual could be competent to exercise such powers, even if self-governed to be the very model impersonal meritocrat. Partiality of knowledge would itself produce biases. So it is questionable whether science could ever be governed impartially by an individual, and whether individuals can ever be fully free of partiality. These may seem to be wholly academic questions, but in understanding the partly marketized forms of current science governance, it is important to realize that they arose out of, and as reforms of, systems that looked very much like this. John Wesley Powell, the head of the United States Geological Survey, the largest scientific organization of the nineteenth century, served as the decision-maker on whose desk, in Truman's famous expression, the buck stopped. In Germany, Friedrich Althoff, the creator of what Max Weber criticized as the Althoff system, exercised similar powers over the whole university system, determining the courses of the careers of academics by determining what positions they would be offered.[1]

The usual solution to the problem of partiality and the problem of conflict of interest is to collectivize decision-making and base it on a consensus of the competent. If the government hands money over to scientists organized in a collective body, who use such mechanisms as peer review to fund those projects that scientists agree to be the most worthy, the government has a mechanism of distribution, a means of assuring some degree of accountability (through the threat of withdrawing funding if scientists acting collectively do not live up to their collective responsibilities). This form of delegation is arguably the most efficient means of reaching at least some state goals, notably the advances of science. When we consider the problem of governing science, what we ordinarily mean is to impose some form of governance that is more restrictive and directive than this sort of governance. This was the solution that in practice took hold in the twentieth century, against which Beck rebels, and which Truman rejected in theory but accepted in practice by the National Science Foundation of the United States as a simple consequence of the impracticality of doing anything different. The actual legislation as it was approved created a presidential board of governors of the foundation that was not self-governed. Opinions differ over whether in practice this amounts to little more than a fiction, but it is an interesting fiction since it brings the National Science Foundation itself into line with a long historical tradition of similar bodies that take the form of commissions. One may then ask why the commission form seems to have such an natural affinity to quasi- science decision-making processes, and go on to think through the significance of "commissions" in a "knowledge society" (which is to say, in the terms I have used here, a society in which quasi-science, decisions requiring science but not settled by science, plays an increasingly large role). I will return later to the problem that concerns Beck, which we can now restate: the concern that experts have *collective* interests or collective biases or partialities that conflict with the interests of those they purport to serve.

Two Kinds of Consensus

Collectivizing decisions, even by such modest means as using peer-review to create artificial competitive situations for the granting of research funds, involves a special use of the concept of "consensus," a use that goes to the heart of the distinction I have made here between quasi-science and science. The core distinction is between consensus on the corpus of basic scientific knowledge, that is to say what scientists agree to be scientifically true, and consensuses of other kinds that require scientific knowledge, but for which science is not sufficient. A simple example of this is to be found in grant applications. Scientists predict results from their work, predict outcomes, and predict potential benefits from their work, and also predict historical outcomes of decisions to fund or not to fund their research (cf. Turner 1999). Decision-making seeks a consensus, or rests on a consensus achieved through procedures of voting or scoring.

From a political point of view these differences are best understood in terms of the difference between the *consensus of scientists* and *scientific consensus*. Scientists themselves distinguish what is and is not a genuine scientific fact.[2] Speaking *as scientists* they may also attempt to construct a consensus of competent scientific peers, not merely with respect to their opinions about the truth of a model but with respect to their opinions about the sufficiency of the evidence and the adequacy of their explanation. With this we are already in a domain of quasi-science, where "science is required but not sufficient." The "meta"-judgments about sufficiency of evidence are *unlike* those made within science about candidates for the corpus of scientific knowledge itself. And it is these judgments that most closely resemble those involved in the governance of science.

Many problems other than funding, of course, are subject to discussions among scientists and attempts are made to establish a consensus with respect to the relevant facts. In attempting to identify a scientific consensus with respect to given problems, the National Institutes of Health regularly hold meetings that include experts of various kinds who are asked to evaluate the existing research, fit it together, iron out and evaluate inconsistencies, and arrive at a policy relevant to scientific conclusions. These are in effect status reports, reports on the state of the evidence with respect to a particular question and of the areas in which the evidence more or less strongly supports some policy relevant conclusion. Which are these: scientific consensuses or consensuses of scientists? Clearly the latter. The consensus that these reports represent is a forced consensus in the sense that it asks for an assessment of the evidence at a particular point in time and with respect to a particular evidential situation.

Judgments about the status of something as an accepted scientific fact, of the candidates for addition to the corpus of scientific knowledge, are different in intent. They are not understood to be time-relevant or evidential situation relevant, though it may well turn out that the evidential- situation changes in an anticipated way that undermines the previous consensus. It is perfectly reasonable to complain that the two kinds of consensus are illegitimately run together in discussions of science as it relates to policy questions, and indeed the distinction is frequently intentionally observed. Attempts to employ standards appropriate to science, such as peer review, to judgments about policy-relevant issues, such as global warming, for example, run together the two kinds of issues, typically without explanation or argument. The fact that claims made about global warming are subjected to "peer review," for example, does not make them into facts like those of core science, even if the reviewers and the review process are the same. Peer review is a procedure of evaluation, a decision-making device.

There is a no need here to construct an artificial, "principled" distinction between core science and quasi-science, for it is a distinction in the world—a distinction made within science, by scientists, on a continuous basis. Roughly, core science consists of the results that are accepted as fact, as belonging—textbooks, as permanent (however differently it might be construed in the future) and accepted or deserving of acceptance as such by, in principle, all scientists. "Peer review" has an elective affinity to the notion of acceptance in "accepted scientific fact," but not a close one. Papers that are accepted for publication by peer review are not, as outsiders wrongly think, "bullet-proofed" by the process, but rather merely accepted as legitimate subjects of discussion—interesting and not obviously false or inadequately supported by experiment.

Peer panels can produce many other things as well, and do, especially in the governance of science: decisions regarding the funding of research or the awarding of prizes, fellowships, graduate student support, and so forth. And they can also be used to produce, or simply legitimate, claims in quasi-science, such as claims about global warming or adolescent pregnancy. These are topics in which methods of modeling are controversial and where there are "facts" that are accepted as scientific, but where these facts alone are far from sufficient to warrant strong predictions, especially about the effects of policy. Typically, they are subject to the infirmity discussed previously in relation to cholera. They do not concern closed systems, in which all parameters are known and all relevant causes included. They represent good guesses, and typically the topics themselves are such that good guesses are the best knowledge that one can have.

In the case of adolescent pregnancy, to choose a behavioral science example, it will never be true that all variables will be known, or that an isolatable basic process will be identified that is subject to control. But policy requires "knowledge." So it is not unreasonable to do what in fact was done: to bring together the main researchers, with their data, for a discussion that was designed to determine what consensus existed, both on the problem and on the state of the evidence. This is a paradigm case of quasi-science.

There may well be perfectly good *political* grounds for adopting such a strategy for obtaining a "consensus of scientists," and it may be perfectly appropriate to apply them in these cases. And there may be good political reasons for making other choices. Standard-setting and regulatory decisions are often made by committees reflect both the representatives of interested parties, scientists familiar generally with what is known about the causal processes and scientific knowledge about the relevant chemicals or disease processes as well as the relevant technologies.

The political question to ask about such committees, and the discourse that occurs within them, is somewhat different from, but not entirely independent of the epistemic question of what it is that these committees know that justifies their decisions. Granting a committee the right or responsibility to set standards, for example, for the concentration of some particular chemical in drinking water, is a political act. It presumes that a committee of a particular kind is in sense a better way of making a decision than other available options, and 'better' may mean a great many things. It may simply be more convenient to delegate such decisions when legislatures feel that they are either not competent to make the decisions or if, for more Machiavellian reasons, they consider it useful to shift the blame for decisions onto another body. Or it may simply be that it is more convenient for representative bodies being lobbied by special-interest groups to remove from their own consideration decisions that cannot be made in ways that these groups regard as beneficial.

The Cholera Case: Nineteenth-Century Expertise and its Political Absorption

So far I have strayed very far from the problem with which I began: the place of expert opinion in liberal democracy. But we are now in a position to see in the abstract what sort of problem this is. The state is in the business of organizing discussion as the basis of state action, and in the business of acting itself. "The State," however, is an abstraction. In practice, the state delegates its executive authority to civil servants, or bureaucracies, and delegates much of the "discussion" as well, to bodies of various kinds, or to civil servants in bureaucracies, who use their discretionary powers to regulate and control, or to induce cooperation. There are strong national difference in the patterns of delegation.

Beck's formulation of the problem of expertise as a problem of cooperation, as I have suggested, is "German," and it recalls the strategies of cartelization and government support that were the basis of the industrial development of Germany in the nineteenth century, and a prominent part of the German *Sonderweg*. The same must be said of the patterns of science funding: The German method involved industrial support for academic laboratories in which industry was induced to "cooperate" by bureaucrats with various broad discretionary powers. These strategies of delegation and organization repeat themselves throughout the history of German science. Such strategies are not imitable where appropriate bureaucratic traditions do not support them. When the United States sought to do so, imitating Japan with DARPA in the 1990s, the results were not impressive. But something can be said about the differences in regimes of handling expertise apart from the lesson that solutions do not travel well, illuminating the problem of governance of science.

In what follows I will consider the example of cholera. In recent writings on cholera in the nineteenth century, scholars have exposed the extent to which the "right" experts were ignored, the distinctiveness and contingency of the situations in which they were, and the extent to which politics, and especially bureaucratic politics, played a significant role in the failure to both receive new ideas about cholera and implement the necessary measures, particularly in the creation of sanitation processes and water filtration. The story of the Hamburg epidemic of 1892 is emblematic of expertise gone wrong. The reactions of the St. James Parish Board (London), which was persuaded by a commission that included John Snow to remove the handle of the Broad Street pump (1854), is emblematic of right decision making. The actions of the sanitary commission of the city of New York also stand out as a success: but so does the complex background to the creation of this board, relevant here because it is a classic "commission." The actions in Britain of the General Health Board, which operated with a bad theory of cholera that was only slowly abandoned, and that out of bureaucratic self regard, once accepted, was claimed credit for, are examples of partial success. London was spared an epidemic like Hamburg's as a result of their efforts.

Richard Evans's classic text on the Hamburg cholera epidemic of 1892 comes to the following conclusion:

> Hamburg experienced a major cholera epidemic in 1892 for three basic reasons. Last in order of importance, and coming into operation only when the other two factors had had their effect, was the chronic overcrowding, poverty, and malnutrition which . . . existed in virtually all the poorer areas of the city, above all after the new harbor construction of the 1880s. This acted as a "Multiplier" of the disease by facilitating its rapid spread from person to person. It could only come into action because the disease was carried to virtually every household in the city by mains water. The failure of the Senate and the Citizens' Assembly to agree on a proper filtration system for the water-supply until it was too late, and the failure to implement a comprehensive system of sewage disposal and treatment, must be accounted the principal reasons for the epidemic proportions reached by the disease. . . . Most important of all was the Hamburg authorities' policy of concealment and delay. (Evans 1987: 304)

This is a good point to begin the comparisons, because this was a case in which expert knowledge—science—was catastrophically misgoverned.

The misrule occurred through a combination of the inability of "public dialogue between the broadest variety of agents"—to use Beck's phrase—to reach agreement, and a powerful official bureaucratic apparatus able to keep secrets. So we may take it not only as a test of Beck's model, but as a test of a certain model of governance of science: one in which powerful bureaucracies are directed and controlled by public discussion that is itself governed by elaborate procedures. The alternatives to this model are those in which intermediate bodies—what David Guston calls boundary organizations, or what I have called here "commissions, operate to provide facts or conclusions that can be made the subject of public discussion."[3]

Beck, curiously, says little about the prosaic sociological problems of the organization of implementation. But one must assume that he has in mind powerful bureaucracies with strongly coercive power: the kinds of regulative powers he wishes to grant to "cooperative" discussion require it. Cooperation presupposes a distribution of powers among those who are doing the cooperating, or requires an agency to enforce the cooperative arrangements. What Beck seems to imagine is even more invasive, and incidentally characteristically German—a situation in which the function of the bureaucracies is to induce cooperation, its power is to a large extent to grant permission to act, and in which the power to withhold permission is decisive and final, not subject to judicial review or much direct legislative control. Envisioning such instruments in the governance of science is difficult, as we shall see— science, as Michael Polanyi (1951) stressed a half century ago, characteristically involves competition and risk-taking and choice under competitive conditions. So understanding how powerful bureaucracies, and their alternatives, such as boundary organizations, have actually functioned in domains that involve science, areas of "quasi-science," is one of the few ways we have of envisioning alternative possibilities for the governance of science.

In Hamburg there had been, prior to the epidemic of 1892, a long political discussion over the problem of drinking water, of precisely the inconclusive kind that Beck supposes should be an expected outcome of what we may call "expert-egalitarian" discussion, by which we may mean "discussions in which everyone is treated as undifferentiated with respect to their expertise." A political decision needed to be made for filtration and clean water, which was costly but, according to the theories accepted elsewhere, essential to avoiding cholera spread through the water supply. The decision was blocked through disagreement over how to charge for it, as well as over skepticism about the need. The *popular* opinion that the available river water was especially pure was an important reason that the political discussion failed to produce agreement. Yet the Hamburg politicians *did* consider expert advice from their local bureaucrats and the local medical community: the two were closely entwined, and supported a theory of cholera derived from Max von Pettenkofer (Rosenberg 1962: 194–7, 199–200).

The board acted in accordance with a process in which they were constrained not by the view s of other experts, which might have forced them to consider alternatives to their own view, but were directly controlled only by the Hamburg senate and lower house, politicians and notables who provided the sole form of official public discussion. Hamburg also had commissions, but public discussion in these commissions conformed rather closely to Beck's ideal: the commission was ignored (Evans 1987: 158). There was no delegation of decision-making power to expert bodies, no "monopolization," which is what Beck in principle rejects. Delay and non-decision were the consequence.

The inadequacy of the advice of the Hamburg medical community reflects a more general phenomenon. One of the features of bureaucracies is that career advancement is heavily dependent on conformity. Strong bureaucracies penetrate into the professional or expert community, affecting the climate of opinion within them. The effect of a powerful bureaucracy in this case was to assure conformity with what turned out to be a mistaken theory of cholera. But powerful bureaucracies of this sort succeeded elsewhere. In Berlin, the same kind of bureaucratic power produced conformity with what turned out to be the right theory, and Berlin was spared. But Koch's powers, as described by Evans, are the fullest realization of the intertwining of bureaucratic power and control of opinion through the control of careers:

> Koch could . . . be assured . . . of vigourous backing from the Imperial government in imposing his views on cholera prevention on medical authority throughout the Empire Already in June 1884 he was made a member of the Prussian Privy Council (*Staatsrat*) and co-opted onto the Cholera Commission for the German Empire. This had hitherto been controlled by Pettenkofer. Koch became the dominant force. In the same year he organized a course of the diagnosis and prevention of cholera, in which 146 doctors took part, including 97 civilian (i.e., non-military) doctors from all parts of Germany and 20 other countries. In 1885 he became full professor (*Ordinarius*) of Hygiene at the University of Berlin, and was appointed Director of a specially created Institute for Infectious Diseases in 1891. These positions enabled him to influence large numbers of pupils in favour of his ideas and methods. His influence was further spread by his senior pupils. . . . Koch founded a journal for the propagation of his ideas, the *Zeitschnft für Infektions-kamkheiten.* Thus Koch and his pupils were rapidly taking over the field of hygiene. (Evans 1987: 266–7)

So the expert advice that the politicians dealt with was in each case essentially monolithic, but different. The expert with authority in Berlin was Koch. But Koch had no direct authority over Hamburg, whose physicians were influenced by Pettenkofer, a Munich physician. Pettenkofer took the view that the cholera bacterium alone could not produce the disease—and in a famous demonstration, he drank a beaker of infected water, to no ill effect—which he thought required other conditions, including "fermentation" in the ground, which implied different public health measures.

If we consider two other cases with different structures, the results are revealing. In London, there was a national bureaucracy, the General Health Board, headed by a statistician and sanitary reformer named William Farr. Farr was wrong about cholera—he was a miasmatist, who had produced an impressive curve fit to data on cholera deaths in London, which he published, based on the idea that elevation decreased the number of cholera deaths in an epidemic. He also produced a vast quantity of research relating to other variables, especially social variables, water quality and so forth, none of which produced the startling consistency of the elevation data. The bureaucracy headed by Farr was never powerful as a regulatory body. Decisions about water, for example, were local. Moreover, the career structure of London medicine was such that no bureaucracy had much ability to assure the conformity of the local medical community. Farr's office had a monopoly on the official publications it produced about cholera, but these were not binding on local authorities. Nor was there any particular career benefit to conformity to the position of the General Health Board. And this led to a different outcome.

Beginning with the political achievement, only now being acknowledged, of the St. James Parish committee: it was persuaded to remove the handle of the Broad Street pump, now one of the most celebrated episodes in the history of medicine (Brody et al. 1999). The means by which this political act occurred are partly known. The Parish committee, faced with a local outbreak of cholera, what we would now call a "cluster," appointed John Snow and Henry Whitehead, a clergyman, to a special cholera inquiry committee to deal with an outbreak of cholera near Golden Square. He applied the spot map technique and found that 73 of the 83 cases were near the Broad Street pump, which Snow reasoned was the source. Whitehead was skeptical of this explanation at first, but was soon convinced. Some who drank the water escaped cholera, others who didn't drink it contracted the disease. But the proportions were overwhelmingly in support of the pump hypothesis. The Parish committee asked Snow to write the report up, and Whitehead himself figured out what had contaminated the pump water—a leaky cesspool three feet from the pump. The material came from the washing of the diapers of an infant who had died and for whom the cause of death was listed as diarrhea. The discovery led the Parish committee to excavate the pump area, which revealed that the brick linings of both the well and the cesspool had decayed, allowing seepage. The pump's handle was removed, and the outbreak subsided. This was an act of a small political body faced with an emergency, but in a position to create its own commission, listen to its conclusions, and act independently on them—or decline to act.

This dramatic episode was only a small part of the story, however. Snow's own efforts began long before this episode, and the absorption of his views continued long after. The medical background was complex. Cholera was the most researched disease of the century, and many correlations, as well as many well-attested cases, were part of its large literature. Snow, a private physician, was struck by the many remarkable cases in which cholera spread over vast distances, apparently carried by individuals, strange cases in which cholera attacked one group, such as the passengers of a particular ship, and spared those that had left from the same port at the same time, and cases where one company of soldiers passing a water hole and drinking from it left healthy, and the next one became deathly ill. These cases were difficult to square with any sort of miasmatic or "fermentation" account. He hypothesized, as it turned out correctly, that the real cause was minute material in the evacuations of the victims that got into the water supply or was otherwise ingested. In an era in which proportionality of cause and effect was a standard methodological rule, and before the microbe account of disease was accepted, this was a radical idea. It was also easy to regard it not as radical, but as old news. Even Farr's research office agreed in some respects with the basic idea: Bad water was one of the many variables they found to be associated with cholera. But bad water was poorly defined, and not defined in a way that was readily amenable to policies that allowed epidemics to be stopped or prevented. In the long run, this was the loophole through which the bureaucracy grudgingly accepted Snow's arguments—as though they had been their own all along. But it was an important loophole, for it allowed for the institution of reforms that had the desired result.

Snow's hypothesis was startlingly reconfirmed as a result of a natural experiment in which a mysterious outbreak of cholera occurred after changes had been made in the water supply, but only among the customers of one water company. This appeared to refute Snow. It was then discovered that the company had been illegally drawing water from a source that was "impure." What is striking about this story is, on the one hand, the obduracy of the bureaucratic experts, though they did eventually concede that Snow was right, and on the other the ability of Snow—and the motivation—to persuade the Parish committee, which promptly created a "commission" rather than attempting to make the decision on its own, to act on his ideas, and the openness of the committee to being persuaded. The result was that London did not have another cholera epidemic after the changes in the water supply were fully effected.

The American version of this story is equally interesting, largely because it represents a different combination of politics, expertise, and bureaucratic structures. In the United States, public health matters in the nineteenth century were for the most part controlled by municipalities, which were, in the case of New York and other major cities, democratic in a particular sense—dominated by the "spoils system."[4] The major device for dealing with the threat of cholera was sanitation, and sanitation contracts were a political plum. Boards of health, in a typical arrangement that applies to many boards even today, were composed of elected officials who were stakeholders of various kinds who sat as the board of health and used its special powers when circumstances required. The politicians—Democrats, in this case—preferred to conduct business as usual. But they were vulnerable to reformers, and in the manner that they were defeated the deep roots of national political and bureaucratic traditions become visible.

Tocqueville, writing a few decades before, had observed that

> A single Englishman will often carry through some great undertaking, whereas Americans form associations for no matter how small a matter. Clearly the former regard association as a power means of action, but the latter seem to think of it as the only one. (1969: 514)

This is precisely how cholera was attacked. As Charles Rosenberg notes in his history of the American response to cholera, "it is hardly surprising that New York's Citizens' Association (an informal group of respectable—and predominantly Republican—Gothamites organized early in the 1860's to promote 'honest government') should sponsor a subsidiary Council of Hygiene and Public Health." (1962: 187). This "council"—a case of a Tocquevillian association—surveyed the sanitation arrangements of the city, and reported the dismal results of the sanitary regime in place. Another arm of the Citizens' Association was at work on reforming die political structure that produced it, proposing a bill in the state legislature to create a board of health that did not operate on the spoils system, and had experts rather than politicians on it. This was a lesson drawn from the examples of Paris and London, but also from Providence, Rhode Island, and Philadelphia. The bill required that the board consist of medical men trained especially for public health work (ibid.: 188–91).

The state of knowledge in 1866 was expressed by the New York Academy of Medicine, yet another Tocquevillian association, which advised the medical profession to

> for all practical purposes, act and advise in accordance with the hypothesis (or the fact) that the cholera diarrhoea and "rice-water discharges" of cholera patients are capable in connection with

well-known localizing conditions, of propagating the cholera poison, and that rigidly enforced precautions should be taken in every case to disinfect or destroy these ejected fluids. (Quoted p. 195)

The resolution reflected some medical politicking—the "local conditions" clause assured unanimity, though few doubted that the discharges alone were the cause. But it also reflected the internationalization of expertise on the topic, and the rapidity of "conversions" to the "Snow-Pettenkofer" account of the disease (which in practice was the Snow account) was impressive. The New York Sanitary Commission, which had been granted enormous power by the legislature, acted accordingly: "the bedding, pillows, old clothing, and utensils—anything that might 'retain or transmit evacuations of the patient'— were piled in an open area and burned" (p. 205). New York City escaped cholera, other states copied the legislation (p. 211), and the contained "epidemic" of 1866 was the last serious cholera threat in the United States.

The politics of opposition are worth noting, especially in light of Beck's demands. The Commission was imposed by the state legislature, which was dominated by Republicans, the Democrats, Catholics, and immigrants of New York City opposed it as the creation of rural lawyers that favored the rich at their expense (p. 208). The New York Sanitary Commission was composed of experts, but "public" experts, rather than individuals who were the creation of consensus producing career structure in a powerful bureaucracy. They were accountable professionally but also accountable as public figures for their actions, in the sense that their reputations were closely tied to the outcomes of very public decisions. And there were associations, such as the Citizens' Association and the New York Academy of Medicine, that were independent watchdogs, with an eye on the practices of other governments and on international expertise. A member of such a board was highly constrained by these bodies—someone who did not wish to jeopardize a carefully built-up reputation both as an individual and as an expert would be obliged to resign or protest bad decisions.

Beck's model contrasts with this rather sharply, because in a situation of expert egalitarianism, reputation is unimportant or equalized, as is responsibility for the outcome: rather than being held responsible, a person can behave irresponsibly without consequences or act in terms of self-interest without consequences. Indeed, this is a major part of the Hamburg story. The issue of taxation that was entirely a matter of political interest prevented the reaching of a resolution, exactly as

Beck says is a permissible outcome, consequently filtration devices were not built until after the epidemic had taught the public lesson that they were necessary. To be sure, of course, had Hamburg been as fortunate as Berlin and had the right leading figures in its bureaucracy, bureaucratic power and the consensus it favored would have produced the right decision. But the issue of "governance" is not eliminated by the existence of powerful bureaucracies. Someone needs to pick the powerful bureaucrats and to judge the bureaucracy. The Hamburg notables and politicians, who were closely related to the medical community, proved incapable of doing so. Thus the combination of interest group democracy and powerful bureaucracy in this case proved fatal, and more generally is prone to the same very particular kind of error.

Governing Science

Germany had the best science at the time—Koch won a Nobel Prize for identifying the cholera bacterium. Yet it had the worst cholera epidemic of Europe, long after other countries had solved their cholera problem. Is it too much to compare this to the situation in German physics during WWII? There again, Germany had the best scientist, Heisenberg, and the best intellectual resource base. The customary view of this episode is that authoritarianism led to failure. Heisenberg, unconnected by vigorous debate from his subordinates, made a key error and failed to see the solution to the problem of fusion. But in a larger perspective, the problem may be one of the organization of scientific activity: The bureaucratic structure of the research effort led, as it led in the Hamburg medical community, to a consensus that turned out to be false. And it is difficult to imagine a powerfully bureaucratic mode of the governance of science that would not be systematically prone to this kind of error.

James Bryant Conant, writing after WWII, described the following model for presenting expert opinions to decision-making bodies (1951: 337–8). To avoid what he took to be the central problem of scientists promoting their own hobby horses, he suggested that even a good idea ought to be able to withstand criticism and that opponents be selected to play the role of the devil's advocate. These opponents would promote alternative proposals so that the decision makers would be given a genuine choice; the experts would be forced to articulate arguments not merely persuasive to nonexperts but tested against the proposal of the expert opponent. This left judging in the hands of nonexperts, but gave experts their due as pleaders of cases. There is a sense in which this model is the one that most closely represents the situation in St. James Parish, and in the commission that the Parish committee created when it joined Snow with a skeptical clergyman. Snow obviously was arguing against something and it should have been known to everyone as it certainly was known to Snow that this was not only a minority view but a view opposed by the official bureaucracy. Snow nevertheless prevailed, producing perhaps the single best decision in the whole cholera affair.

The New York model is also revealing: expertise constrained by expert scrutiny, where the "outside" experts are genuinely independent, and the reputations of the experts exercising authority are, in effect, marketized, so that they would suffer for their obduracy, constrains experts very effectively, while at the same time producing decisive results—the New York methods were highly effective and easy to imitate. Whether this is a model that can be used in other political traditions, such as the German, in which "cooperation" is the working norm, is open to question. But in each case there is some means of protection against the error-prone combination of bureaucratic power and the quasi-scientific "consensus of scientists."

Liberal democracy does not require the leveling of expertise, as Beck assumes. Decisions and fact-finding on matters of quasi-science can be delegated, and delegation, in the form of commissions, is a standard device. But it does require that the consensuses produced by these delegated discussions are reasonably free from conflicts of interest and partiality—something effectively precluded by powerful bureaucracies that create climates of opinion.

Notes

1. I will leave aside for this discussion an aspect of the political status of these figures that elsewhere (e.g., 1986 and 2001) I have stressed—the fact that they purported to act as representatives. Powell explicitly saw himself as representative of both the scientific community and of the democratic public; Althoff presumably saw himself as a civil servant, exercising the authority of the state.

2. All of the usual qualifications apply to this distinction. Results that appear to be part of the corpus may cease to be so on the basis of future scientific investigation. Moreover, the distinction is often hazy in practice. The difference between adding to the corpus of scientific knowledge and performing statistical analysis using scientific concepts, or coming up with findings that are potentially relevant to the corpus of scientific knowledge but not yet sufficiently understood to be part of that corpus, for example, is not always clear or acknowledged by scientists, who may loosely use the term "scientific." And investigations that employ the categories and concepts of

the corpus of scientific knowledge without themselves being candidates for inclusion in this corpus, are not entirely and obviously distinct m practice from "basic" investigations. Nevertheless, there is a difference, however disputed in practice, that is acknowledged by scientists and forms part of their working self-concept and scheme of evaluation.

3. Elsewhere I have argued for an extended use of this term to include self-created bodies with claims to represent larger bodies of opinion, which can have a variety of roles in relation to state arrangements (cf. Turner 2002). This reflects the political reality that international NGOs such as Greenpeace play a role similar to official commissions in public discourse, and take on the same kinds of trappings of expertise.

4. The Tammany Hall bosses did not have the benefit of Robert Merton's later analysis of them as benign helpers to impoverished immigrants, and, as we shall see shortly, for the most part stood in the way of "reforms" that kept these same immigrants from dying of cholera. The distribution of contracts and jobs was their concern, and doubtless some of this benefited the poor that the bosses protected. But the cost/benefit ratio of this arrangement was not very good, and in the case of cholera, its exploitative character was obvious. But so was its "Democratic" character, for most of the public health measures that were called for, such as the removal of pigs from the houses of citizens, were unpopular with the poor.

References

Beck, Ulrich (1992) *Risk Society: Towards a new modernity.* Mark Ritter (trans.). London: Sage Publications.

―――― (1994) "The reinvention of politics: Towards a theory of reflexive modernization," in *Reflexive Modernization*, Ulrich Beck, Anthony Giddens, and Scott Lash (eds.). Stanford, Cal.: Stanford University Press. 1–55.

―――― (1995) *Ecological Enlightenment: Essays on the Politics of the Risk Society*, Mark Ritter (trans.). Atlantic Highlands, N.J.: Humanities Press.

―――― (1997) *The Reinvention of Politics: Rethinking Modernity in the Global Social Order.* Mark Ritter (trans.). Cambridge, U.K.: Polity Press.

Conant, James (1951) *Science and Common Sense.* New Haven, Conn.: Yale University Press.

Dicey, Albert Venn ([1914] 1962) *Lectures on the Relations between Law & Public Opinion in England During the Nineteenth Century.* Second edition. London: Macmillan.

Evans, Richard J. (1987) *Death in Hamburg: Society and politics in the cholera years 1830–1910.* Oxford: Clarendon Press.

Habermas, Jürgen (1973) *Theory and Practice,* John Viertel (trans.). Boston, Mass.: Beacon Press.

Jasanoff, Sheila (1995) *Science at the Bar: Law, science, and technology in America.* Cambridge. Mass.: Harvard University Press.

Penick, James L. Jr. (1965) *The Politics of American Science: 1939 to the Present.* Chicago: Rand McNally.

Peters, Richard (1956) *Hobbes.* Harmonsworth, England: Penguin Books.

Polanyi, Michael (1951) *The Logic of Liberty: Reflections and Rejoinders.* Chicago: University of Chicago Press.

Rosenberg, Charles E. (1962) *The Cholera Years: The United States in 1832, 1849, and 1866.* Chicago: University of Chicago Press.

Scott, Roy V. (1970) *The Reluctant Farmer: The rise of agricultural extension to 1914.* Urbana, Ill.: University of Illinois Press.

Turner, Stephen P. (1999) "Does Funding Produce its Effects? The Rockefeller case," in *The Development of the Social Sciences in the United States and Canada: The role of philanthropy*, Theresa Richardson and Donald Fisher (eds.). Stamford, Conn.: Ablex Publishing. 213–27.

Wynne, Brian (1996) "May the Sheep Safely Graze? A reflexive view of the expert-lay knowledge divide," in *Risk, Environment, and Modernity: Towards a new ecology*, Scott Lash, Bronislaw Szersynski, and Brian Wynne (eds.). London: Sage Publications. 44–83.

Concluding Observations

Free Flow of Information or Embedded Expertise? Notes on the Regulation of Knowledge

Reiner Grundmann

Jeffrey Klein has frankly described two different approaches to the problem of *Governance of Knowledge*. He thinks that there is a German approach in that "some of the German scholars . . . fear a chaotic future because they are planners." Americans, on the other hand, "want to bring on the future as fast as they can because they are competitive optimists" (Klein, *infra*). Likewise, Stephen Turner has pointed to the fact that "Beck's formulation of the problem of expertise as a problem of cooperation . . . is 'German.'" Turner goes even further, putting Beck in a historical context that "recalls the strategics of cartelization and government support that were the basis of the industrial development of Germany in the nineteenth century, and a prominent part of the German *Sonderweg*" (Turner, *infra*). What is more, science funding in Germany follows a specific model, involving "industrial support for academic laboratories in which industry [is] induced to 'cooperate' by bureaucrats with various broad discretionary powers. These strategies of delegation and organization repeat themselves throughout the history of German science" (ibid).

While there is much to recommend such a reading, we also should bear in mind that other contributors to this volume do share the view that regulation of knowledge is necessary. And Fukuyama, in his recent book on biotechnology, addressed the question of how to react to the challenges posed by biotechnology in the following way: "The answer is obvious: *We should use the power of the state to regulate it.* And if this proves to be beyond the power of any individual nation-state to regulate, it needs to be regulated on an international basis." (Fukuyama 2002: 10, orig. emph.).

In what follows, I do not intend to analyze the papers presented in this volume in order to find traces of national characters. However. I will point occasionally to some telling differences in political culture and intellectual approach where appropriate. Let us take the sections in turn.

The Separation of Knowledge and Ethics: How to Combine Them?

In Part I on the Emergence of Knowledge Politics and its Origins, two papers are of German origin, one British. Gernot Böhme thinks that "the necessity of regulating the generation and the application of knowledge in this context is incontestable. The question is only at what point, and through whom, this regulation should occur." Böhme identifies two connected problems. The first is that knowledge and ethics are separate and therefore any attempt at regulation has to come from outside of science. The second is that such regulation from outside practically always comes too late. Böhme thus explores the question of "whether the regulation of knowledge might not be a task for scientists themselves, and whether the presence of ethical points of view ought not to be implicit in their approach to their object" (Böhme. *infra*).

He imagines that a more holistic science could emerge provided its *curative promises* were comparable to the established life sciences. If it could show that it provides similar possibilities at a lower price compared to the manipulative life sciences and biotechnologies, then it would have real chance. Böhme is optimistic: "This does not seem so hopeless. An analogy may make this clear: had we invested from the beginning in alternative forms of energy, especially solar energy, the same means as were applied to fostering nuclear energy, then today the alternative forms of energy would long since have become competitive. In order to create an analogous situation for the life sciences and thereby for a future medicine, to be sure, a *mobilization from within*—that is, a forceful movement among the scientists themselves—would be required." (ibid.). Such a mobilization could occur as a product of public pressure.

Wolfgang van den Daele in his paper shares the basic point of departure in that he sees modern science and ethics separated. However, he is much less optimistic compared to Böhme. He sees this as an irreversible rift between science and ethics which cannot to be remedied in the foreseeable future. "Even if scientists could be persuaded (and sanctioned) to observe a new Hippocratic oath that they provide knowledge only for the sake of humanity and the common good (or sustainable development, for that matter), such an oath would be an external control on the application of knowledge not an intrinsic feature of their knowledge" (van den Daele, *infra*). Thereby the essential tension between knowledge and ethics would merely reaffirm itself.

Steve Fuller comments on both authors and finds that "Böhme's and van den Daele's proposals for governing knowledge underestimate the normative import of institutions. Both are primarily oriented to articulating a normative vision—Böhme's of an updated *Naturphilosophie* and van den Daele's of a state-enforced cultural pluralism—that leave open the exact institutions that might realize them." This is even more important as Fuller feels a need to justify the very regulation, as this might be seen as authoritarian, "if exclusive attention is paid to articulating the ends of knowledge production, thereby implying that virtually any means is justified in the process of achieving those ends" (Fuller, *infra*).

Which institutions does Fuller propose? In short, he tries to rescue the university from its present predicament as either diploma mill or provider of vocational training. Taking up on Böhme's and van den Daele's point of departure, he gives this a special twist when he says that "questions of the governance of knowledge dwell in a knowledge-society split at least as old and problematic as the divide between mind and body." Firstly, Fuller states that universities are entrepreneurs following Schumpeter's formula of "creative destruction" of social capital, a.k.a. knowledge. "By virtue of their dual role as producers and distributors of knowledge, universities are engaged in an endless cycle of creating and destroying 'social capital', that is, the comparative advantage that a group or network enjoys by virtue of its collective capacity to act on a form of knowledge (Stehr 1994). Thus, as researchers, academics create social capital because intellectual innovation necessarily begins life as an elite product available only to those on 'the cutting edge.' However, as teachers, academics destroy social capital by making the innovation publicly available, thereby diminishing whatever advantage was originally afforded to those on the cutting edge." (Fuller, *infra*).

At first sight, there seems to be a curious convergence between Fuller and van den Daele in that both discount insights from the sociology of science that stress the implicit character of knowledge, that is, the claim that knowledge cannot be reduced to information but is always embedded in social contexts and communities of scientific practitioners (e.g., Fleck, Kuhn, Polanyi). Thus Fuller states that innovators in science need not "necessarily have been in direct contact with the researchers whose work theirs builds upon or, for that matter, overturns. Rather, these innovators encounter their precursors secondhand, through textbooks and their often undistinguished classroom interpreters." And van den Daele goes even further when making the claim that while traditional knowledge is local, community-related and personal, modem scientific knowledge can be conceived of as information and is global, thus valid across community boundaries and impersonal.

But Fuller ultimately rejects such a view and believes that knowledge is more than textbook information. This is only logical given his focus on institutional forms of knowledge production. Would knowledge only be information, institutions would be less important as it would flow freely. Institutions would be important only if knowledge would flow too freely and thus create disincentives for knowledge providers—a point made by Edmund Kitch—or if it would lead to undesirable side effects (a point to which I shall return below). Discussing Kitch's argument that we need protection of intellectual property rights in order to protect the innovators from the threat of copycats, Fuller states "saying that knowledge tends to escape from its original bearers is a far cry from saying that knowledge tends to be made universally available. The former is simply a long-winded acknowledgement that communication exists; the latter implies that all communication is mass communication. *However, the fact that knowledge is not a private good does not mean that it must therefore be a public good.* Kitch's vision comes dangerously close to this non sequitur, which overlooks the importance of institutions in the construction of knowledge as a public good" (Fuller, *infra,* orig. emph.).

Here it would be interesting to take on board the relevant literature in greater detail. As Paul David and others have argued, knowledge is a commodity *and* a public good; it can be institutionalized under the three different regimes of patronage, public procurement and private property ("the three Ps," David 1993; see also Foray 1999). Fuller's point that the university has a special institutional place is well taken. And he is right to point out that we need institutional analysis when talking about the governance of knowledge. However, his recommendation of the university as occupying center position seems too narrow since also this institutional form can cover "the three Ps."

Functional Differentiation as German *Sonderweg*?

In Part II, Major Social Institutions and Knowledge Politics is the topic. Here we have three papers of which one is of German origin. Werner Rammert takes as his point of departure the theory of social differentiation à la Luhmann and distinguishes three types of differentiation in society (to which he then adds a fourth): the segmented, the stratified and the functional. Rammert proposes that there exists a close relationship between specific types of social differentiation and a corresponding regime of knowledge production. To be sure, he does not define this as a causal relation, but supposes "that particular patterns of dominant social structures favor certain means of coordination and suggest specific institutional answers" (Rammert, *infra*).

More specifically, in a segmented society (made up of clans, families and tribes) one will find the same stock of knowledge in every social unit. In a stratified society, there is a clear vertical or center-periphery distinction of knowledge and social status. Knowledge becomes centralized and systematized at the top or the center and distributed from there to lower (or peripheral) parts in the social order. In modern society however, we see a functional principle of differentiation at work, or so the theory suggests. Several social subsystems are horizontally separated from each other and enjoy relative autonomy (it is a disputed issue among theorists of social differentiation how far this autonomy reaches, with some claiming that social systems are autopoietic, i.e., self-producing and self-reproducing systems that are closed in their operations).

None of these systems has priority or command over any other. So how is knowledge production performed in this type of society? One would expect that the scientific subsystem is the system where knowledge is produced. However, Rammert points out that "scientific knowledge production . . . gains high institutional autonomy and self-governance; but in order to exploit it for the sake of economic innovation or military power enforcement, markets of patents and licenses and big organizations were recruited to coordinate the separated, but complementary innovative activities. A *regime of complementary and specialized knowledge production* belongs to the functional type of social differentiation" (Rammert *infra*). In other words, science may be functionally separate from other systems, but its application requires the coordination of several social systems.

At this point he introduces a fourth type of social differentiation, the fragmental type. Examples for such fragmented social entities are industrial districts (or innovation networks). Fragmentation in this definition shares an important feature of the segmented type, that is, to combine different types of knowledge, but it differs from it in that "its mix of heterogeneous knowledge consists of fragments of a once systematized and functionally specialized knowledge. Elements of all kinds of knowledge are recombined. The fragmental differentiation differs from the functional one radically under the aspect that the purified separation is given up in favor of heterogeneity and reflexivity. Functionally specialized institutions and purified scientific disciplines remain important factors on the back stage of fragmented society, but they are losing their privilege to act on the front stage where network forms of organization and transdisciplinary epistemic and expert cultures take over the prominent roles. The *regime of heterogeneously distributed knowledge production* rises in close relation to the type of fragmental" (Rammert, *infra*).

With this theoretical innovation, Rammert tries to reconcile a "typical German" trait of social theory (functional differentiation of society) with recent approaches in various fields of social science inquiry, from "mode 2 knowledge production" though Neo-Institutionalism to Actor Network Theory, to name but a few. If there is a German *Sonderweg*, he attempts to make it connect to a larger road network.

Catastrophic and Moral Risks

William Leiss, in his essay on Policing Society: Genetics, Nanotechnology, Robotics, starts by pointing out that anyone who seeks to challenge the great founding faiths of modern society—that of the infinite benefits of the liberation of the natural sciences from the intellectual and institutional shackles of dogma, including religion—is in for a rough ride. Reference is made to our background paper (Grundmann and Stehr 2001), but there are others who advocate similar positions regarding the question of social surveillance and regulation of knowledge (e.g., Fukuyama 2002). And Leiss himself comes to accept such a position. He arrives at it through an examination of the likely consequences of technological applications such as rDNA and nanotechnology. In his view, a new category of risks is created by these, which he calls moral risks. As their negative and evil aspects are virtually unlimited, they are also catastrophic risks.

> I define "catastrophic risk" in this sense as the possibility of harms to humans and other entities that call into question the future viability

of existing animal species, including our own. Thus these are not only risks to the present generations of living animal species, but also to future (perhaps *all* future) generations of presently existing species . . .

The catastrophic risk[s] . . . stem from current research programs that are widely distributed around the world; moreover, the strongest drivers of them are private corporations, including the large pharmaceutical multinationals, acting with lull encouragement, support, and incentives from national governments. Especially where the possible health benefits of genetic manipulations are concerned, the combined public-private interests are overwhelmingly supportive, driving the research ahead at an accelerating pace. Governments especially are enthralled with the economic significance of these new technologies, are competing with each other under innovation agendas to capture major shares of the corporate invertments, and are loathe to stop and think about unintended consequences. (Leiss, *infra*).

The question arises if "our institutional structures, including international conventions, are robust enough to be able to contain such risks within acceptable limits; or alternatively whether these risks themselves should be regarded as unacceptable, a position that would impel us to seek to forbid individuals and nations from acquiring and disseminating the knowledge upon which those technologies are based" (Leiss, *infra*).

When posing the question of regulation, Leiss suggests that "it is extremely difficult even to think about controlling either the process or the results." The reason he gives harks back to our earlier discussion about the explicit or implicit character of knowledge. The same discussion arises when it comes to technical applications. In his contribution, Rammert follows recent work in the sociology and history of science and technology studies when pointing out that modern science incorporates more and more technical instruments. As a consequence, "implicit skills and explicit knowledge are closely interwoven with the 'epistemological objects.' Harry Collins (1974) had already demonstrated that the published explicit physical and technological knowledge is not sufficient, if one wants to replicate a successful experiment at another scientific laboratory; the replication must include a person who was a member of the scientific research group or who shared the practices of the group for a while as a visitor" (Rammert, *infra*).

If we are concerned about catastrophic potential of scientific and technical artifacts, we get an interesting contrast between a more optimistic view which sees knowledge (including its technical applications) as implicit, localized, and personal, and a pessimistic view which sees knowledge basically as free floating. Why is the first view optimistic, the second pessimistic? The first view is optimistic (at least in its implications) because it assumes that the widespread availability of pure information does not create the necessary and sufficient conditions for technical applications that work in practice. The second view is pessimistic (at least in its implications) because it assumes that the widespread availability of pure information will in fact create the necessary and sufficient conditions for technical applications that work in practice. Given that all scholars involved in this debate would like to prevent catastrophic risks, those who see knowledge as free floating information should have the strongest wish to regulate new, potentially dangerous knowledge.

Leiss, however, does not assume that knowledge is free floating, yet advocates strong regulations. This points to a further possibility of diffusion of knowledge. It could be that there are more and more specialists around who can handle the knowledge and its potentially disastrous applications. Leiss argues along these lines when saying that "knowledge is widely distributed among individual scientists [and] . . . among private actors (corporations) which have the option of moving their operations on a regular basis, seeking perhaps the least-regulatory-intensive national base on the globe." Not only are more and more people knowledgeable of the relevant techniques, "the technologies themselves become increasingly 'simplified' and thus easier to hide, if necessary; the genetics technologies, for example, can be carried out in small laboratories almost anywhere" (Leiss, *infra*).

Still, at least for the time being, we might take some small comfort in the fact that despite the free availability of information (through the World Wide Web) of, say, how to make an atomic bomb, efforts by interested (presumably terrorist) groups have failed. And the origin of the recent Anthrax attacks seems to be scientists working in government labs, not Al Quaida. The 'know-how' requires close connections to practitioners of such technologies. And, arguably, government labs are easier to control than terrorist groupings.

Jeffrey Klein argues that there will be little or no or weak regulation of knowledge because "traditional mediating forces seem inadequate, if not complicit. Our species is poised—via technology, science, industry, government and the media—to transform itself" (Klein, *infra*). This is because we all take part in the game of out competing each other and when it comes to the possibilities of making our offspring better off, we all will embrace biotechnology. "But the market for optimizing human nature will fundamentally be run by capitalist and libertarian rules. Science will take a back seat to commerce, and social science may be left far behind in its own foggy idealism" (Klein, *infra*).

Echoing a theme initiated by Fuller, Klein asks "Could the universities act as a brake on such Faustian ambitions?" He thinks this to be unlikely. "World-class research universities like Stanford try to stoke the ambitions of the Jim Clarks [who made a $150 million gift to Stanford to fund a new bioinformatics department) in order to benefit from their largesse. . . . Even if successful entrepreneurs weren't calling the shots, would-be entrepreneurs on the faculty will protest any strictures on them" (Klein, *infra*).

And circling back to the character of knowledge he states that "the Internet refrain that 'Information wants to be free' could be revised to say. 'Information seeks transmission.' We are communicative creatures." In this version it becomes obvious that the view of knowledge as information is largely ideological. Maybe this has more to do with the fact that Klein had a previous life as investigative journalist than with the fact that he is American. His succinct prose seems to seduce him to make statements which are too quick, as when he states that "technology in general and biotech in particular promise to give consumers what they desire. Governance structures in the knowledge market will remain weak because neither producers nor consumers want to slow down, and competitive pressures won't let them, anyway." But exactly this remains to be seen. After all, it was the United States that led international efforts at environmental and health protection through the 1970s and 1980s. And if it is true that Germans (or Europeans for that matter) "fear a chaotic future because they are planners" and thus set in motion attempts towards effective regulations, Americans like Klein might find allies in them and see the dynamics of capitalism less as a destiny and more as a product of human intervention and social shaping.

Patients, Laypersons, and Consumer Power

Let us move on to Part III and examine several case studies on the governance of knowledge. These are studies on the contested use and usefulness of the concept of race (Troy Duster), on the BSE saga in the United Kingdom and its institutional aftermath as manifested through the government's response (Kevin Jones), and on the public discourse on genetically modified (GM) foods (Javier Lezaun).

Troy Duster opens this section with a compelling account of how the concept of race is being resurrected for practical clinical purposes "while the full range of analysts, commentators and scientists—from post-modern essayists to molecular geneticists to social anthropologists—have been busily pronouncing 'the death of race'" (Duster, *infra*).

The interesting point is that the concept of race is socially meaningful, without doubt more meaningful than definitions based on DNA or RNA. As Duster puts it., "Even though people may someday come to understand that they are basically similar at the level of the DNA, RNA, immunological or kinds of blood systems-it is the language group, kinship, religion, region and race that are still far more likely to generate their pledge of allegiance" (Duster, *infra*).

"Race" is used by a wide variety of lay people, giving rise to all kinds of folk beliefs. In order to "correct" such beliefs, the American Anthropological Association, for example, states that "physical variations in the human species have no meaning except the social ones that humans put on them" (quoted in Duster, *infra*). This suggests that there are two levels of reality, one based on a scientific description of differences among humans based on concepts like DNA, and a popular one based on 'mere' social constructions.

However, as Duster is quick to point out: "There is profound misunderstanding of the implications of a 'social constructivist' notion of social phenomena. How humans identify themselves, whether in religious or ethnic or racial or aesthetic terms, influences their subsequent behavior. Places of worship are socially constructed with human variations of meaning and interpretation and use very much in mind. Whether a cathedral or mosque, a synagogue or Shinto temple, those 'constructions' are no less 'real' because one has accounted for and documented the social forces at play that resulted in such a wide variety of 'socially constructed' places of worship."

This line of thought is applied to the issue of race. Duster argues that race as social construction "can and does have a substantial effect on how people behave." In other words, social constructions can be more powerful in their effects compared to "objective" knowledge—as for example in accounts based on DNA screening and sampling.[1] Duster is interested in possible feedback loops between behavior and the biological functioning of the body. Modifying the well-known aphorism of W.I. Thomas, he states: "*If humans define situations as real, they can and often do have real biological and social consequences*" (Duster, *infra*). This is evident, e.g., "African American males, for example, may by identifying as African Americans, be more likely to eat a category of food ('soul food') that might systematically put them at higher risk for prostate cancer."

Another example is the degree of darkness of skin. As dark pigmentation is associated with hypertension (at least in the United States). Studies have found that darker skin color is a good predictor of hypertension among blacks of low socioeconomic status. However, this is not true for blacks of any shade who are well employed or better educated. What is more, "poor blacks with darker skin color experience greater hypertension 'not for genetic reasons' but because darker skin color subjects them to greater discrimination, with consequenty greater stress and psychological/medical." While this does not establish nature of the causal relationship. But when "practicing physicians see 'darker skin color,' their diagnostic interpretation and their therapeutic recommendations are systematically affected." They are likely to make systematically different recommendations for treatment of heart disorders, by race, even when patients have the same presenting symptoms.

Kevin Jones ("BSE and the Phillips' Report: A Cautionary Tale about Government uptake of risks") studies the results of from the BSE inquiry which was submitted to the British House of Commons in October 2000 and became known as the Phillips Report. The task of the inquiry was to review the emergence and identification of BSE and vCJD in the United Kingdom, along with the adequacy of the actions taken by the Government in response to the disease. Jones pays specific attention to the report's use of the concepts of risk and uncertainty. He also addresses the strengths and weaknesses of the Phillips Report and the government response to the report.

Jones holds that Beck and Giddens' theory of risk society have influenced the Phillips report. One indication is that the relationship between New Labour's presentation of Third Way politics and Giddens' proclamations about the nature of late modern British society are clearly interwoven. However, he is careful not to overstate this connection: "I do not wish to argue that Phillips has full-heartedly embraced ideas of the risk society in reporting the outcomes of the BSE Inquiry. However, terms such as 'risk' and 'uncertainty' become overarching themes throughout the report, and although Beck and Giddens in no way hold a monopoly on the use of the terms, Phillips does appear to tentatively approach several of the key tenets of their theories."

The most important point here is that both Beck and Phillips think that society must address the fallibility and inconclusiveness of science. If knowledge is uncertain (as many scholars in the sociology of science maintain), we cannot assume that scientists have the right answers, are able to establish cause-effect relationships or scientific proof. They are therefore not in a position to protect us from the potential dangers we face in our everyday lives. Secondly, proponents of the risk society argue that the hazards facing society today are the unpredictable consequences of human actions and not the result of nature's power.[2] Our efforts at dominating nature have produced unintended side-effects with which we must deal (the efforts of which are likely to create ever new side-effects). Within this context of risk and epistemological uncertainty, traditional social institutions are no longer able to cope with the protective tasks which they have been assigned and on which society depends.

According to Jones, "Phillips is decisive in stating that the human health tragedy of BSE was not directly attributable to the uncertainly of the science of BSE and vCJD, but to the Government's mishandling of this uncertainty. By relying on a series of false assumptions they produced an unstable and dangerous platform on which they made, or failed to make, the regulatory decisions necessary to avoid disaster." The main failure pertains to the many reassuring statements which were issued by the government and its expert committees.

So how does Jones evaluate the Phillips report? He thinks that the fundamental error the Phillips Report made, was to treat the idea of risk as an inevitable object, or certainty. Risk management strategies prioritize "the management of the *consequences* of uncertainty. As a result, he displays little regard to the potential for reducing the a priori social production of risk." For example, Phillips clearly lays blame on "intensive and industrialized agricultural systems" as the source of BSE: "BSE developed into an epidemic as a consequence of an intensive farming practice 'the recycling of animal protein in ruminant feed . . . instead of addressing agriculture as a social practice, Phillips obsesses over the minute technological, scientific, and industrial processes behind the risks associated with BSE. One is hard-pressed to find a discussion of agriculture itself, and is instead presented with page after page outlining the particularities involved in the slaughter, rendering and distributive use of an animal" (Jones, *infra*).[3]

Furthermore, it appears that "although the report takes a positive step in promoting an open and transparent model of communication, it is limited by the adoption of conservative assumptions about the nature of knowledge and expertise." Jones maintains that "at the center of the Phillips Report's approach to the communication of risk and uncertainty is an assumption that trust can be generated by simply improving communicative processes and developing an awareness of the rationality of the public at large." However, science is perceived as the *exclusive procurer* of knowledge, while the public is seen as *passive receivers*. Laypersons may have been in a better position to ask awkward questions such as: How do you know? And: why not?

The lay public and their involvement in risk controversies is the topic of Javier Lezaun's contribution (Subjects of Knowledge: Epistemologies of the Consumer in the GM Food Debate). Lezaun tells the fascinating story of how the U.K. food industry tried to introduce GM food into the British market. Key player is the Institute of Grocery Distribution (IGD) which produced research for the British food suppliers. In this role it was instrumental for unifying the strategics and interests of the different actors of the food chain, from farmers to retailers. It latched on to a European Commission proposal to install "informed consumer choice" as the keystone of institutional control of GMOs. The enlightened consumer with the "right to know" was taking center stage in official pronouncements from EU institutions, consumer, and environmental groups. Based on Raymond Williams, Lezaun draws attention to the fact that "manufacture is not only related to the supply of known needs but to the planning of given kinds and quantities of production which required large investment at an early and often predictive stage . . . the creation of needs and wants and of particular ways of satisfying them, as distinct from and in addition to the notification of available supply which had been the main earlier function of *advertising*" (as quoted in Lezaun, *infra*).

Lezaun's essay shows how the British food industry in the early 1990s developed a sophisticated strategy to introduce biotechnology foods into the British market. Labeling of genetically modified products was key to these efforts. "By the end of the decade, however, no products labeled as 'genetically modified' could be found on British supermarkets' shelves, and most retailers and suppliers were in the process of phasing modified ingredients from their products." How was this possible? The answer is, the strategy of labeling backfired. In the early 1990s, the IGD undertook a large research effort in order to "understand the customer." In 1996, the first GM product, Zeneca's tomato purée, was on sale in supermarkets. The cans of purée were clearly labeled as "produced from genetically modified tomato," and were sold at a cheaper price than similar cans of conventional tomato purée. Due to the extensive PR of this product. Sainsbury's and Safeway sold 1.6 million cans. This was clearly a commercial success.

However, as the British public became very concerned about GM food in early 1999, the reliance on this success turned out to be fatal.[4] The IGD undertook a new round of consumer research in March 1999. The result was that there had been no significant change in the profile of the U.K. consumer. The scapegoat was quickly identified in media exaggeration of the risks. But this did not help in dealing with the labeling issue. Any GMO label now had become a clear liability. Lezaun draws our attention to the fact that

> as the food industry reflects on their failure to introduce genetically modified products, and subtly revisits its assumptions about the public of biotechnology, the epistemology of the citizen-consumer it produced over the last decade remains alive and well in the policy-making bodies of the European Union. However, this subject of consumption has proved rather unmanageable. As consumer and environmental groups have shown, informed choice and consumer rights are a powerful tool to oppose the market release of GM foods—if not the instrument of societal control of the technology. Food consumption is now a site of political contention and social contestation. As the food industry discovered, and the European institutions may soon find out, governing the unruly consumer they are helping create is not an easy task (Lezaun, *infra*).

Differences in Political Culture

In the last chapter, we have contributions from Schulte, Turner and Lessig, dealing with Issues in Knowledge Politics as a New Political Field. Schulte presents an analysis of the German legal system which is facing the problem of scientific uncertainty. His theoretical reference point is Luhmann's theory of social differentiation and autopoeitic closure of communication. He claims that experts are required in legal decision making in order to absorb uncertainty. However, and paradoxically, they produce new uncertainties through their very activity. The important point from his theoretical perspective is that legal decisions can only lead to other legal decisions, not to resolve scientific and/or technical disputes around risk issues.

How can we reconcile the ideals of democratic decision making with the need of scientific expertise? This question is tackled in Turner's essay. The tension that exists since expert knowledge is inaccessible to non-experts, "who are forced, like it or not, to accept the authority of experts." According to Turner, it would be a mistake to think that the creation and acceptance of science's privileged or monopolistic status is inherently anti-democratic. The question comes down to the question of delegation. Provided that there is little disagreement over the necessity to do so, the question is what to delegate and how to delegate. The necessity exists as we cannot assume that technical competence and specialized knowledge lies with members of parliament or with generalist civil servants (although one could make the point that important tasks of modern bureaucracies are derived exactly from this need). One can therefore imagine various ways of delegation—"to representatives directly, to "the people," to some sort of 'cooperative' arrangement, or to some sort of specialized body" (Turner, *infra*).

Turner's suggests a solution of "some novel kind of forum in which experts could be in some sense held accountable or forced to legitimate themselves." Based on various historical examples, he points out the perils of leaving too much room for lay knowledge which may be ill informed or manipulated by short-sighted interests, as was the case in the Hamburg cholera epidemic of 1892. The Hamburg notables and politicians were too closely related to the medical community. "The bureaucratic structure of the research effort led, as it led in the Hamburg medical community, to a consensus that turned out to be false. Thus the combination of interest group democracy and powerful bureaucracy in this case proved fatal," Turner concludes. His way forward draws upon J.B. Conant's model who believed that

> even a good idea ought to be able to withstand criticism and suggested that opponents be selected to play the role of the devil's advocate or to promote alternative different proposals in order that the decision makers would be given a genuine choice but also that the experts would be forced to articulate arguments that would be not merely persuasive to nonexperts but tested against the claims of the expert opponent of the scientists with a proposal. This left judging in the hands of nonexperts, but gave experts their due as pleaders of cases. (Turner, *infra*)

Of course, this relates to the different policy styles in the United States and Europe. While the United States follows an adversary style, Europe is driven largely by a consensus style. This difference has important implications for transatlantic debates, as Turner remarks:

> Whether this is a model that can be used in other political traditions, such as the German, in which "cooperation" is the working norm, is open to question. But in each case some means of protection against the error-prone combination of bureaucratic power and the quasi-scientific "consensus of scientists" [would be needed] Liberal

democracy does not require the leveling of expertise, as Beck assumes. Decisions and fact finding on matters of quasi-science can be delegated, and delegation, in the form of commissions, is a standard device. But it does require that the consensuses produced by these delegated discussions are reasonably free from conflicts of interest and partiality. And this is something that powerful bureaucracies which create climates of opinion effectively preclude.[5]

Lawrence Lessig, last but not least, turns our attention to the nature of the Internet as a common pool resource. Drawing on the work of Elinor Ostrom, he argues that their importance in a wide range of contexts across the world has been demonstrated, including such diverse settings as communal tenure systems in Switzerland and Japan to irrigation communities within the Philippines (Ostrom 1990). However, in the U.S. intellectual culture things are viewed differently, as Lessig points out: Here, "a commons is treated as an imperfect resource. It is the object of 'tragedy,' as biologist Garrett Hardin famously described; it provides an imperfect incentive to consume and deploy resources. Wherever a commons exists, the aim is to enclose it. Commons are unnecessary holdovers from time gone buy, best removed if possible" (Lessig, *infra*).

This underlines the point made by Klein which I quoted at the beginning. He highlighted the difference between "German scholars who are afraid of a chaotic future because they are planners," and Americans who "want to bring on the future as fast as they can because they are competitive optimists." Applied to the case of the Internet, the reign of possessive individualism distinguishes the two cultures. And Lessig basically agrees:

"For most resources, for most of the time, this bias against the commons makes good sense. It is true that when some resources are left in common, individuals are driven to over consume the resource and therefore deplete it. But for some resources, this bias against the commons is blinding. Some resources are not subject to the 'tragedy of the commons,' as obviously, some resources cannot be 'depleted.'" Ostrom's point is that the common pool resources she investigated can in fact be depleted, but are not. The fact that they are not depleted in every case requires careful examination. The Internet, by contrast, is more like a radio signal, a scientific theory or a poem which cannot be depleted by frequent use. But no matter how misplaced the rhetoric of the advocates of privatization is—Ostrom's model does not apply here as the problem is not one of resource depletion but of privatization of a commons which is working well—Lessig shows the general suspicion towards common resources which is prevalent in the United States.

To sum up. There seem to be two sticking points in the debate. One has to do with the characterization of knowledge as either information which flows freely among actors and across rime and space or as socially embedded knowledge which is dependent upon communities of practitioners. Depending upon which view one adopts, this has implications for regulation, especially for regulation of risky technologies. The second point pertains to key actors of regulating knowledge. Various possibilities have been examined in this volume, including scientists and experts, laypersons, and consumers. Also institutional forms have been examined, and universities, networks of distributed knowledge and public judgment of expert debates have been proposed. Of course, there are no definitive solutions to be expected in this exciting new field of research which the governance of knowledge promises to be. But some important markers have been put up which will inform our future research efforts.

An Anglo-German Debate?

A final note on the difference between U.S. and European approaches to regulation in general and to the governance of knowledge in particular. First of all, it may be futile to try and identify such general differences. Even Turner's thorough critique of Beck's theory does not reject it on general grounds. The points he raises are examined in great detail, and an alternative institutional form is proposed which tries to capture much of Beck's initial motives. Second, I have tried to separate the cultural-political orientations of the contributors to this volume from the prevalent orientations among the wider political culture. For it is one thing to state that there is such a difference but it is another matter to reproduce and reinforce the difference from an academic viewpoint. The fact that scholars working in a U.S. context have raised the issue does not necessarily mean that they are supportive of their own political culture. The same applies for European scholars, and it may be difficult for all to escape such cultural viewpoints entirely.

Should we adopt an "American" optimistic or a "European" pessimistic viewpoint? All depends on how we judge the above key issues. If we believe that the catastrophic potential of new knowledge based technologies is rather large, or if we believe that knowledge is free floating and available to anyone who wants to apply it in technical ways, or if we believe that communities are not closed enough in order to retain the virulent knowledge, then we should be rather cautious towards such developments and favor strong regulations. This aspect is very much concerned with the abuse of knowledge based technical possibilities.

There is a second aspect. Here we deal with the "normal" working of new technologies—technologies which may eventually turn out to produce unintended side-effects. In this case we need a different orientation, one that takes into account differences in lay and experts' judgment and various forms of institutionalization. Various authors of this volume have presented their views on this and pointed to possible solutions (ranging from the development of scientific ethics, bureaucratic control, establishing networks, making legal decisions, explore the role of the universities). In conclusion, I would like to draw the attention to something slightly different. Both aspects from above are connected in that lay perceptions will give more weight to maximum damage, not statistical probabilities of its occurrence (Kahneman et al. 1982). Depending on how an issue is presented to lay persons, we will get different responses. If we want to establish precautionary policies, we would need to include the lay public in decision making, at the same time depicting the catastrophic potential prominently (via the mass media). This will produce a strong risk averse view on the side of the public. Opponents of such a course of action will try to do exactly the opposite, that is, exclude the public where possible, point to the benefits of the technologies and/or the low probability of something going wrong. Whatever decision is ultimately made, it will be legitimized by a reference to neutral scientific expertise. Welcome to the politics of knowledge!

Notes

1. This is only the case if we were to share the premise of this argument, i.e., that scientific knowledge is outside of social constructions. There are good pounds to challenge such a view and thus the argument would be recast in a way in which the social construction of race proves more robust than the social construction of DNA.

2. The Phillips Report states: "In a primitive society, the major hazards are those posed by nature. In a complex modern society the acts of individuals or corporate bodies may also involve serious hazards to other members of society" (Phillips et al. 2000: 31). Luhmann and Beck wrote similar things in the mid-1980s.

3. One could argue that the attention to specific detail is the strength of the report and that a general critique of intensive agriculture would have been perceived as waffling. Additionally, the fact that Lord Phillips is lawyer indicates that one of the latent functions of the Inquiry was to establish a verdict about liability. In fact, no individuals have been singled out in this

respect. This sets an important precedent for possible liability claims in the future.

4. There are various reasons for this, among them the research conducted by Arpad Pusztai at the Rowett Institute in Aberdeen and publicized by the media in August 1998. After his summary dismissal, numerous scientists expressed their support for his work in February 1999 thereby reigniting the controversy (see Grundmann 2002).

5. Governments may also select expert commissions which will propose what their preferred course of action is, thereby lending scientific authority to their decisions. Shils put it this way: "Advisors are too frequently chosen not so much because the legislators and officials want advice as because they want apparently authoritative support for the policies they propose to follow. It is obvious that in complying with these desires, the legislators and the officials are in collusion with the scientists to exploit the prestige that scientists have acquired for objectivity and disinterestedness" (Shils 1987: 201).

References

David, Paul A. (1993) Knowledge, property and the system dynamics of technological change. *Proceedings of the World Bank Annual Conference on Development Economics 1992,* Washington, D.C.: World Bank.

Foray, Domique (1999) Science, Technology and the market, *World Social Science Report.* London: UNESCO Publishing/Elsevier.

Fukuyama, Francis (2002) *Our Posthuman Future: Consequences of the biotechnology revolution.* London: Profile.

Grundmann, Reiner (2002) Advocate scientists and their role in risk controversies: The cases of CFCs and GM foods, Ms.

Grundmann, Reiner, and Nico Stehr (2001) Policing Knowledge: A new political field. Kulturwissenschaftliches Institut, Essen, Germany, Ms.

Kahnemann, D., P. Slovic, and A. Tversky (1982) *Judgement under Uncertainty, Heuristics and Biases,* Cambridge: Cambridge University Press.

Ostrom, Elinor (1990) *Governing the Commons: The evolution of institutions for collective action.* New York: Cambridge University Press.

Phillips, Lord of Worth Matravers, June Bridgeman, and Malcolm Ferguson-Smith, 2000. *The BSE Inquiry Volume I, Findings and Conclusions.* London: The Stationary Office.

Shils, Edward (1987) "Science and scientists in the public arena," *The American Scholar* 35: 185–202.

Stehr, Nico (1994) *Knowledge Societies.* London: Sage.

Contributors

Gernot Böhme is professor emeritus of philosophy of the Technsiche Universität Darmstadt, Germany. He studied physics, mathematics, and philosophy. From 1969 to 1977, he was a research fellow at the Max-Planck-Institute, Starnberg (Habermas, v. Weizsäcker). His main fields of interest are classical philosophy, social studies of science, philosophical anthropology, philosophy of nature, theory of time, aesthetics, ethics, and Goethe. Böhme has published 30 books and more than 200 articles. Among his recent publications in English *The Knowledge Society* (co-editcd with Nico Stehr, 1986); *Coping with Science* (1992); and *Ethics in Context: The Art of Dealing with Serious Questions* (2001).

Wolfgang van den Daele is director of the Standard-Setting and Environment research unit at the Social Science Research Center Berlin (WZB), and professor of sociology at the Free University of Berlin. From 1985 to 1987 he was a member of the German Federal Parliament's Select Committee on the Opportunities and Risks of Genetic Engineering. At present, van den Daele is a member of the National Board of Ethics for the Federal Republic of Germany. Recent publications include *Gruene Gentechnik im Widerstreit* (with Alfred Püler and Herbert Sukopp, 1996); *Kommunikation und Entscheidung* (with Friedhelm Neidhardt, 1996); "Von moralischer Kommunikation zur Kommunikation über Moral. Reflexive Distanz in diskursiven Verfahren," *Zeitschrift für Soziologie*, 30 (2001), and "Das Subjekt als Grenze der Techniksteuerung? Gewissen, Angst und radikale Reform— Wie starke Ansprüche an die Technikpolitik in diskursiven Arenen schwach werden," (in *Politik und Technik: Analysen zum Verhältnis von technologischem, politischem und staatlichem Wandel am Anfang des 21. Jahrhunderts*, 2001).

Troy Duster is professor of sociology at New York University, and has an appointment as Chancellor's professor at the University of California, Berkeley. He is chair of the Board of Directors of the Association of American Colleges and Universities, and a member of the National Advisory Board of the Social Science Research Council. From 1996 to 1999, he served as a member of the National Advisory Council for Human Genome Research, and during the same period served as member and then chair of the joint NIH/DOE advisory committee on ethical, legal and social issues in the Human Genome Project (*The ELSI Working Group*). His books and monographs include *The Legislation of Morality* (1970), *Aims and Control of the Universities* (1974), *Cultural Perspectives on Biological Knowledge* (co-edited with Karen Garrett, 1984), and *Backdoor to Eugenics* (1990), a book on the social implications of the new technologies in molecular biology and *Race: Essays on the Concept and its Uses in Multi-Racial and Multi-Cultural Societies* (1995). His most recent publications on this topic are "The Sociology of Science and the Revolution in Molecular Biology" (in *The Blackwell Companion to Sociology*, 2001); "Social Side Effects of the New Human Molecular Genetic Diagnostics" (in *The Genomic Revolution: Unveiling the Unity of Life*, 2002); and "The Social Consequences of Genetic Disclosure" (in *Culture and Biology*, 1999).

Steve Fuller is Auguste Comte Professor of Social Epistemology in the Department of Sociology at the University of Warwick, UK. Originally trained in history and philosophy of science, Fuller is best known for his foundational work in the field of 'social epistemology,' which is the name of a quarterly journal that he founded in 1987 as well as the first of his more than twenty books. He has recently completed a trilogy relating to the idea of a 'post-' or 'trans-' human future (or 'Humanity 2.0'). His latest book is *Knowledge: The Philosophical Quest in History* (2015). His works have been translated into over twenty languages.

Reiner Grundmann is a sociologist and political scientist by training. He obtained his Ph. D., in political and social sciences from the European University Institute, Florence. His main research interests are in social and political theory, and the sociology of science, technology, risk, and the environment. He was with the Wissenschaftszentrum in Berlin and the Max Planck Institute for the Study of Societies (Cologne). Currently he is a senior lecturer at Aston University, Birmingham (U.K.). He is the author of *Marxism and Ecology* (1991) and *Transnational Environmental Policy: Reconstructing Ozone* (2001). He has published *inter alia* in the *British Journal of Sociology, Journal of Classical Sociology, Social Science Information, International Environmental Affairs, Science as Culture, European Journal of Communication,*

New Left Review, and *Political Theory*. Recently he co-edited works of Werner Sombart.

J. Rogers Hollingsworth is professor of Sociology and History and Chairperson of the Program in Comparative History at the University of Wisconsin. Awarded honorary degrees by Uppsala University (Sweden) and by Emory University, he is the author or editor of numerous books and articles on comparative political economy. One of his major research interests is the study of how organizational and institutional factors influence different types of innovations. His publications include *Contemporary Capitalism: The Embeddedness of Institutions* (with Robert Boyer, 1997); *Governing Capitalist Economies* (with Philippe Schmitter and Wolfgang Streeck, 1994); *The Governance of the American Economy* (with John Campbell and Leon Lindberg, 1991); and *Major Discoveries, Creativity, and the Dynamics of Science* (with Ellen Jane Hollingsworth, 2011). He is past president and honorary fellow of the Society for the Advancement of Socio-Economics.

Kevin Edson Jones is currently finishing a Ph.D. at the Centre for Research into Innovation, Culture, and Technology at Brunel University in West London. His research investigates how controversies in agriculture and the biosciences relate to normative and political contestations over nature, democracy, and governance in British society. He has also written on the subjects of knowledge, communication and expertise in government organizations.

Jeffrey Klein, one of the founding editors of *Mother Jones*, worked at the magazine from 1976 to 1981, then returned to assume the top editorial spot from 1992 to 1998. During the interim he served as editor-in-chief *of San Francisco* magazine, then founded and ran a Sunday magazine at Knight-Ridder's San Jose *Mercury News*. Klein's op-eds have appeared in many newspapers including the *Los Angeles Times, Newsday, San Francisco Chronicle*, the Portland *Oregonian, Charlotte Observer, Arizona Republic* and the *Guardian* (London). He has been a frequent guest on C-SPAN, Fox News, CNN, and CNBC. He has also served as a guest host on KQED radio's *Forum*, the Bay Area's NPR outlet. Klein is a graduate of Columbia University (B.A. '69) and San Francisco State University (M.A. creative writing '73). He has taught at Stanford's Graduate School of Communications, the University of California at Berkeley's Graduate School of Journalism, and at San Francisco State University. He is the author of *The Black*

Hole Affair, a fact-based thriller about Star Wars weapons that initially ran in serial installments in the Sunday Mercury News. Recently Klein started and sold a software company. He is currently at work on a book about American capitalism.

William Leiss is a fellow and past-president (1999–2001) of the Royal Society of Canada; professor, School of Policy Studies, Queen's University, Kingston, Ontario (since 1994); NSERC/SSHRC industry research chair in risk communication and public policy, Haskayne School of Business, University of Calgary (1999–2004); and executive-in-residence. McLaughlin Centre for Population Health Risk Assessment, University of Ottawa. He is author or co-author of *In the Chamber of Risks: Understanding Risk Controversies* (2001), *Mad Cows and Mother's Milk: The Perils of Poor Risk Communication* (1997), *Risk and Responsibility* (1994), *The Domination of Nature* (1972), *The Limits to Satisfaction* (1976), *and Under Technology's Thumb* (1990); *Social Communication in Advertising* (Routledge, 1990); and *C. B. Macpherson* (1988).

Lawrence Lessig is professor of law at Stanford Law School, and writes in the areas of constitutional law and the law of cyberspace. He is author of *The Future of Ideas* (2001) and *Code and Other Laws of Cyberspace* (1999). From 1997 to 2000, he was Berkman professor of law at Harvard Law School.

Javier Lezaun is currently James Martin Lecturer in Science and Technology Governance and associate professor in the School of Anthropology and Museum Ethnography at the University of Oxford.

Werner Rammert is professor of Sociology and Technology Studies at the Technical University of Berlin; co-founder of the yearbooks *Technology and Society* (82–99) and of the national research program Socionics, member of the Social Shaping of Technology management committee of the European Commission COST A 4. Research projects on the institutional and cultural shaping of technologies (telephone; computer use at home; knowledge-based systems); on the "distributed action" in hybrid socio-technical constellations (multi-agent systems; closed-circuit television; high-tech operation rooms), on constructive technology assessment and on networks of innovation. Latest books (in German): *Technography: A Microsociological Approach to Technology* (co-editor, 2006); *Technology – Action – Knowledge: A Perspective from Pragmatism* (2007).

Nico Stehr is Karl Mannheim Professor of Cultural Studies at the Zeppelin University, Friedrichshafen, Germany. He is a fellow of the Royal Society (Canada) and a fellow of the European Academy of Sciences and Arts. His research interests center on the transformation of modern societies into knowledge societies and developments associated with this transformation in different major social institutions of modern society (e.g. science, politics, governance, the economy, inequality and globalization); in addition, his research interests concern the societal consequences of climate change. He is one of the authors of the *Hartwell Paper* on climate policy. Among his recent book publications are: *Biotechnology: Between Commerce and Civil Society* (2004); *Knowledge* (with Reiner Grundmann, 2005), *Moral Markets* (2008), *Who owns Knowledge: Knowledge and the Law* (with Bernd Weiler, 2008), Knowledge *and Democracy* (2008), *Society* (with Reiner Grundmann, 2009), *Climate and Society* (with Hans von Storch, 2010), *Experts: The Knowledge and Power of Expertise* (with Reiner Grundmann, 2011), *The Power of Scientific Knowledge* (with Reiner Grundmann, 2012) and *Information, Power, and Democracy* (2016).

Martin Schulte is professor of law at the Technical University of Dresden, Germany. He is also director of the Institute for the Law of Technology and Environmental Law at the Law Faculty of the Technical University of Dresden and director of the Center for Interdisciplinary Research of Technology at the Faculty of Philosophy. Schulte has held visiting appointments at the Catholic University of Nijmegen, Netherlands (1989–91) and the Emory University Atlanta, United States (1998). Among his recent publications are "Zum Umgang mit Wissen, Nichtwissen und Unsicherem Wissen im Recht dargestellt am Beispiel des MSE- und MKS-Konflikts" (in *Wissen, Nichtwissen, Unsicheres Wissen*, 2002) and "Wissensgenerierung in Recht und Rechtswissenschaft Selbstbeschreibung und Fremdbeschreibung des Rechtssystems" (in *Szenarien der Wissensgesellschaft*, 2002).

Stephen Turner is graduate research professor and chair of the Department of Philosophy at the University of South Florida. He has written extensively on aspects of the governance of science and the politics of expertise, as well as on the social sciences and their relations with foundations, and reform movements. These writings include "The Survey in Nineteenth-Century American Geology: The Evolution of a Form of Patronage," (*Minerva*, 1987), "Forms of Patronage," in Susan

Cozzens and Thomas F. Gieryn, eds., *Theories of Science in Society* (1990), and *The Impossible Science: An Institutional Analysis of American Sociology*, (Sage, 1990). His most recent book, *Liberal Democracy 3.0: Civil Society in an Age of Experts* (Sage, 2002), examines the conflict between expertise and liberal democratic discourse.

Index

For Product Safety Concerns and Information please contact our EU
representative GPSR@taylorandfrancis.com
Taylor & Francis Verlag GmbH, Kaufingerstraße 24, 80331 München, Germany

www.ingramcontent.com/pod-product-compliance
Lightning Source LLC
Chambersburg PA
CBHW050627280326
41932CB00015B/2550